Kinetics and Thermodynamics of Fast Particles in Solids

Kinetics and Thermodynamics of Fast Particles in Solids

Yurii Kashlev

CISP

CRC Press
Taylor & Francis Group
Boca Raton London New York

CRC Press is an imprint of the
Taylor & Francis Group, an **informa** business

CRC Press
Taylor & Francis Group
6000 Broken Sound Parkway NW, Suite 300
Boca Raton, FL 33487-2742

First issued in paperback 2019

© 2013 by Taylor & Francis Group, LLC
CRC Press is an imprint of Taylor & Francis Group, an Informa business

No claim to original U.S. Government works

ISBN-13: 978-1-4665-8009-1 (hbk)
ISBN-13: 978-0-367-38080-9 (pbk)

Visit the Taylor & Francis Web site at
http://www.taylorandfrancis.com

and the CRC Press Web site at
http://www.crcpress.com

Contents

Preface

This monograph examines the kinetics and non-equilibrium statistical thermodynamics of fast charged particles moving in crystals in different modes. This line of research is very different from traditional ways of constructing a theory of radiation effects, which gives a purely mechanistic interpretation of particle motion. Indeed, the mechanistic theory is limited to taking into account various mechanisms of collisions, while the theory presented in the book, in addition to the collision, takes into account the thermodynamic forces due to separation of the thermodynamic parameters of the subsystem of particles ('hot' atoms) on the parameters of the thermostat (electrons and lattice).

Fast charged particles are described on the basis of statistical ensembles for non-equilibrium systems. This approach has no counterpart in the scientific literature. Generalized transport equations and non-linear equations of relaxation kinetics are used in a wide range of thermodynamic conditions. Generalized equations represent the starting point for further investigation of non-equilibrium processes in crystals in the presence of radiation. Here are some specific objectives:

- Construction of a local kinetic equation of Boltzmann type for fast particles interacting with the conduction electrons and lattice vibrations, on the basis of the principles of Bogolyubov's kinetic theory. The local transport equations are used to calculate the kinetic coefficients of the subsystem of channeled particles, and the possible modes of motion of fast particles in crystals – channeling, quasi-channeling and chaotic motion – are considered.
- Calculation of the equilibrium energy and angular distributions of fast particles at a depth of the order of coherence length, and study of the evolution of particle distribution with increasing depth of penetration of the beam.
- Calculation of transverse quasi-temperature of channeled particles with the heating of the beam in the process of diffusion of particles in the space of transverse energies, as well as cooling

it through a dissipative process. Measurement of the transverse quasi-temperature along the height of the maximum of the angular distribution obtained in the 'shoot-through' experiment

- Research in the framework of non-equilibrium thermodynamics of the relaxation kinetics of random particles, including the thermodynamics of positronium atoms moving in insulators under laser irradiation. Study of the quasi-temperature of positronium atoms.

- Analysis of the kinetics of hot carriers in semiconductors and thermalization of hot carriers, as well as the calculation of the statistical distribution of ejected atoms formed during the displacement cascade. Investigation of the thermalization of the cascade particles in metals and the thermodynamic stability of the subsystem of knock-on atoms.

The book presents the results of recent years, the author's work with minimum information taken from other monographic publications. It sets a new direction of the theory of radiation effects in solids – non-equilibrium statistical thermodynamics of fast particles.

The author tried to focus the reader's attention to this direction and identify possible trends.

1

Theoretical background and model presentation

1.1. Two theoretical disciplines: non-equilibrium statistical mechanics and non-equilibrium statistical thermodynamics

In this monograph two theoretical disciplines are combined as part of the problem of motion of fast particles in crystalline solids, namely, non-equilibrium statistical mechanics and non-equilibrium statistical thermodynamics. These disciplines are connected by deep internal logic as their central problem – the theory of evolution of dynamic systems – is the same.

Non-equilibrium statistical theory (mechanics and thermodynamics) is atomistic science. At its foundation lay dynamic models and statistical assumptions about the probabilities of states at the microscopic description. As for the description of time evolution, both components of the theory use time as an independent variable because time is included in the basic equation of motion – Liouville equation.

Non-equilibrium statistical mechanics is the mechanics of large ensembles, and its main task is to investigate the temporal development of the whole system based on the behavior of its constituent subsystems (molecules of a rarefied gas, the atoms of the crystal, electrons, etc.). It is only important to ensure that the selected subsystems belong to a class that can be described on the basis of the laws of either classical or quantum mechanics. We restrict ourselves to only one class – fast atomic particles moving in solids with velocities exceeding the thermal velocity (although the transition to thermal velocities is also studied.)

We begin by considering the stochastic mechanics of particle motion in the plane (or axial) channels and the mechanics of Brownian motion, which is the traditional division of the stochastic theory. The basic concepts and principles of the theory of random processes are discussed. It is established that the stochastic approach is essential for understanding the majority of the observed orientation effects. One of the main conclusions, which results in the study of Brownian motion and the motion of atomic particles, is that the fluctuations are closely related to irreversible processes, and correlation functions of density fluctuations and fluxes determine the value of the transfer coefficients in these theories.

There are several ways to describe irreversible processes in macroscopic systems. Firstly, the use of phenomenological equations of hydrodynamics, i.e. equations that are not invariant under time reversal, and, secondly, the use of kinetic equations and, above all, of the Boltzmann equation. An important property of the latter is that it predicts the tendency of a non-equilibrium system to equilibrium under very general initial conditions. For example, in the case of a rarefied gas the Boltzmann kinetic equation describes the monotonic approach to equilibrium.

Highlighted is the discussion of the physical picture of Boltzmann, which is given in connection with the development of the local theory of orientation effects. It is based on the ideas of Bogolyubov's kinetic theory [1]. Namely, the Bogolyubov–Born–Green–Kirkwood–Yvon (BBGKY chain) equations were used to obtain local classical kinetic equations for a two-component system consisting of a rarefied gas of the fast particles and a crystal in thermodynamic equilibrium. On the basis of these equations analysis was carried out of kinetic processes in the interaction of particles with thermal vibrations of atoms in the lattice and conduction electrons, as well as an analysis of possible regimes of motion of particles in the planar channels.

The local generalized Boltzmann equation was derived on the basis of a chain of equations and conditions of correlation weakening for the distribution of states of a particle. The collision integral of the Boltzmann equation is used to derive a matrix of random effects in the correlated interactions, with the latter being transformed into a function of dynamic friction and diffusion function in the space of transverse energies. It is shown that these functions have features (maximum, inflection) that are associated with differences in the contribution to the kinetics which comes from particles that move in the channeling mode, quasi-channeling and chaotic motion of the theory. The connection between the three modes of motion of fast particles in deep penetration ($\geq 10^4$ Å) with three modes of motion at shallow depths ($\leq 10^4$ Å) and the characteristics of the backscattering of particles (including the peculiarities of the angular distribution of exit of the backscattered particles).

An essential aspect of the theory of irreversible orientation processes as well as the theory of irreversible processes in general, lies in the fact that the approximate evolution equations are valid only for large times, more precisely, for times exceeding the relaxation time of the system. In this case we have a coarse-grained, smoothed in time picture of evolution in which small details of motion of the particles are blurred, in other words, spread out on a rough time line. This makes it possible to assume that the irreversibility is introduced in this theory by temporal smoothing.

The physical picture of Boltzmann is far from pure mechanics. This can be said at least because the *H*-theorem, obtained using the kinetic equation, interprets the approach of gas molecules to equilibrium as an irreversible process. Then, the equations of mechanics are invariant under time reversal. In fact, the derivation of the Boltzmann equation occupies some intermediate position between the stochastic and dynamic rigorous theory.

Non-equilibrium thermodynamics, based on a stochastic interpretation of the effect of collisions in the Boltzmann picture, has now been formulated. Taking into account fluctuations of the theory near equilibrium leads to conclusions that are in qualitative agreement with the conclusions of Onsager's stochastic description (1931).

We now proceed to discuss the second part – the non-equilibrium statistical theory considered in this monograph, i.e. to discuss the main provisions of non-equilibrium statistical thermodynamics.

Development of the thermodynamics of non-equilibrium processes actually began with the work of Onsager (1931). Onsager is credited with creating the theory of an equilibrium statistical ensemble, using the basic concepts of statistical physics (including fluctuations) for the development of a thermodynamic theory applicable to a wide class of non-equilibrium systems. The essential difference between Onsager and Boltzmann approaches should be emphasized. Indeed, if the Boltzmann equation requires molecular collisions, the Onsager theory does not even consider the very concept of the molecule.

Non-equilibrium thermodynamics of systems, weakly deviating from thermodynamic equilibrium, is based on three assumptions: a linear dependence of the average flows of thermodynamic forces, symmetry relations for the kinetic (Onsager) coefficients and Prigogine's theory of minimal entropy production. We will return to the first assumption, and here we briefly explain the second and third of these assumptions.

The theory which gave the reciprocity relations for kinetic coefficients, characteristic for a number of irreversible processes, was developed by Onsager. Initially, the theoretical conclusions were obtained on the example of the decay of fluctuations. The symmetry relations were then proved in terms of energy transfer in a conductor. As regards the principle of minimum of entropy, according to Prigogine's theory (1947), the steady state of a thermodynamic system is a state in which the rate of entropy has a minimum value compatible with external constraints. In the case where there are no constraints, the system goes into a state in which the indicated speed is zero (a state of thermodynamic equilibrium).

Consider processes in systems deviated from equilibrium. For the macroscopic description of such processes, the system is divided into small elementary volumes, which nevertheless contain a large number of particles. In this case, a quantitative description of the non-equilibrium process is reduced to the compilation of the balance equations for the elementary volumes with the laws of conservation of mass, energy, momentum, and also the entropy balance equation taken into account. Methods of thermodynamics can be used to develop a theory of general principles, without examining in detail the interaction of particles and obtain a complete system of transfer equations.

In order to formulate the law of conservation of energy in the elementary volumes, it is sufficient to take into account that the total energy density consists of kinetic and potential energy and also the density of the thermal motion of particles (internal energy density). The balance equation for internal energy leads to the conclusion that the internal energy is not conserved, but the total energy is, and this includes all of the total energy components. This is known as the first law of thermodynamics.

The second law of thermodynamics expresses the balance of entropy $S(t)$ in the case of entropy in the system formed due to irreversible processes. The equation for the entropy flux $\tilde{J}_s$ includes the heat flux density and diffusion flux density. In addition, the local entropy production depends on the gradients of the thermodynamic parameters. The positivity of the intensity of the source of entropy, which characterizes the local production per unit volume and in unit time, is known as the law of entropy increase in non-equilibrium thermodynamics (the second law of thermodynamics).

The formulation of second law of thermodynamics is based on the inherent definition of irreversibility, since the law allows one to divide all processes to reversible, in which the entropy remains constant, and irreversible, in which the entropy increases with time. The final local entropy production is caused by the processes of heat conduction, diffusion, viscosity, etc., i.e. irreversible processes.

In non-equilibrium thermodynamics, flows in the condition of a small deviation from equilibrium are linearly dependent on the thermodynamic forces and are described by phenomenological equations of the form

$$J_i = \sum_k L_{ik} X_k. \qquad (1.1)$$

Here L_{ik} are the kinetic coefficients (more precisely, Onsager kinetic coefficients). In the so-called direct processes, the thermodynamic force X_k causes the flux J_k, since only $L_{ii} > 0$, and the remaining coefficients are zero. The coefficients L_{ii} are proportional to the transfer coefficient and also such values as the coefficients of thermal conductivity, diffusion, viscosity. In addition to direct processes there are also irreversible processes (thermal diffusion, the diffusion thermal effect, etc.). Cross processes correspond to $L_{ik} > 0$, where $i \neq k$. In view of (1.1), the entropy production and the second law of thermodynamics can be written as

$$\frac{dS}{dt} = \sum_k J_k X_k = \sum_{ik} X_i L_{ik} X_k \geq 0.$$

So far we have considered systems that deviate only slightly from thermodynamic equilibrium. Later, however, attention will also be given to strongly non-equilibrium systems, including the channeled particles in crystals and the recoil atoms formed in a cascade process. These systems, either at the initial moment, or due to processes with finite speed, are characterized by significant deviations from Maxwell's distribution of forward speed (longitudinal) of movement, as well as by deviations from the Boltzmann populations of the levels of transverse motion. Quite often, such a system is characterized by a physical situation where particles of various subsystems that make up the whole system have different quasi-equilibrium distribution and different population levels. In this case, for each subsystem we introduce their individual thermodynamic parameters $F_{im}(t)$, where i is the number of the subsystem.

The importance of the kinetic aspect in analysis of the properties of atomic particles moving in solids does not require any special justification. Spatial and temporal changes in the subsystems of the particles can be caused by relaxation processes and also by temporal evolution of the thermodynamic parameters. The latter circumstance is typical for non-equilibrium thermodynamics. Relaxation processes will be considered further below. As for time changes in $F_{im}(t)$, this change of thermodynamic parameters in the microscopic theory of dissipative fluxes is controlled by $J_{nj}(t)$.

From the above considerations it becomes clear that the subject of research in non-equilibrium statistical thermodynamics is the thermal motion in specific conditions. The specifics of these situations are defined by the presence of uncompensated fluxes of mass $\tilde{J}_k$, heat $\tilde{J}_q$, entropy $\tilde{J}_s$, dissipative flows $\tilde{J}_{nj}$ etc. Non-equilibrium thermodynamics considers such situations both at the phenomenological and the microscopic level of description.

Many equations of non-equilibrium thermodynamics are a generalization of the balance equations for the average particle number density, energy and momentum. The evolution of these variables is composed of slow regular changes associated with coordinated movement of the particles, which corresponds to the macroscopic laws of conservation of energy and momentum of the particles, and fast random changes – fluctuations. Establishing linkages with a wide range of thermodynamic parameters between the flows and the transfer rate coefficients, as well as description of the kinetic coefficients in terms of the basic parameters of the theory, including thermodynamic, is the subject of non-equilibrium statistical thermodynamics.

1.2. Short description of the non-equilibrium system

Stages of the process

In the classical case, the challenge is to derive the basic equations of non-equilibrium thermodynamics, based on the Liouville equation for $\sigma(q, p, t)$ – probability density in phase space (q, p).

$$\frac{\partial}{\partial t}\sigma + iL\sigma = 0. \tag{1.2}$$

Here L is the Liouville operator in classical statistical mechanics, which is determined by the classical Poisson bracket

$$iL\sigma = -\{H,\sigma\} = \sum_k \left(\frac{\partial H}{\partial \mathbf{p}_k} \cdot \frac{\partial \sigma}{\partial \mathbf{q}_k} - \frac{\partial H}{\partial \mathbf{q}_k} \cdot \frac{\partial \sigma}{\partial \mathbf{p}_k} \right), \tag{1.3}$$

where H is the Hamiltonian; $\mathbf{q}_k$ and $\mathbf{p}_k$ are the coordinates and momenta of the particles. In general, the phase space (q, p) is $6N$-dimensional, N is the number of particles.

In the quantum case, the Liouville equation for the statistical operator $\rho(t)$ is written in the same form as in classical theory (1.2), but the Liouville operator is expressed through the quantum Poisson bracket

$$iL\rho(t) = -\frac{1}{i}[H,\rho],$$

where $[A, B] = AB - BA$.

In non-equilibrium statistical mechanics, the Liouville operator (1.3) plays quite the same role as the Hamiltonian in ordinary mechanics. In ordinary mechanics, the Hamiltonian defines the law of evolution of the state of the system, in non-equilibrium statistical mechanics, where the state is represented by the density in phase space, the definition of the Liouville operator completely determines the evolution law $\sigma(q, p, t)$. The difference between the two descriptions is only the determination of what is called the state of the system, but not in the law of evolution.

The most significant feature of the problem is that it is required to determine irreversible in time transfer equations from the reversible Liouville equation (1.2). To obtain irreversible equations from reversible ones, it is natural to abandon the completeness of description inherent in the density $\sigma(q, p, t)$ or in the statistical operator $\rho(t)$, and proceed to an abbreviated description of non-equilibrium states, based on more limited information. In particular, in the classical case for low-density gas, one can use the abbreviated description on the basis of the one-particle distribution function that characterizes the state of one particle of a gas. It is enough to integrate the density $\sigma(\mathbf{q}_1, \mathbf{p}_1, \mathbf{q}_2, \mathbf{p}_2,..., \mathbf{q}_N, \mathbf{p}_N)$ with respect to $d\mathbf{q}_2$ $d\mathbf{p}_2,...,d\mathbf{q}_N\,d\mathbf{p}_N$.

The transition to an abbreviated description of the non-equilibrium state can be achieved by averaging the distribution function with respect to the phase variables or time. As an example, the behavior of a system of oscillators, weakly interacting with a large system – the thermostat – will be considered. Applying the methods of perturbation theory, one can show that in a dynamic system that is affected by a large system there may be stochastic processes (Krylov, 1939).

Derivation of the Fokker–Planck equation in the case of Brownian motion of particles in the environment can be realized on the basis of a chain of equations for the N-particle distribution functions. It is enough to average the equations for the distribution function of a Brownian particle in a macroscopically small interval of time (but still large enough in scale of the characteristic microscopic time). Under this approach, the expression for the friction coefficient of a Brownian particle is derived on the basis of the time correlation function of forces acting on it. Similar formulas, one of the most significant achievements of the theory of non-equilibrium processes, later became known as the Kubo formula, as in [2, 3], they were derived for the most general case of non-equilibrium processes. In [2, 3] there are expressions for the majority of the transfer coefficients via time correlation functions of fluxes, which greatly stimulated the further development of the theory of non-equilibrium processes and non-equilibrium thermodynamics in particular.

The idea, which will be used in the construction of the basic equations of equilibrium theory and the non-equilibrium statistical operator, is based on Bogolyubov' idea of the existence a hierarchy of relaxation times and the possibility of reducing the description of the non-equilibrium process depending on the time scales we are interested in.

In the non-equilibrium processes, depending on the time scale, we can consider various stages that require different descriptions. The existence of these stages is most easily explained by the example of the average density of gas. In this case, the collision time is much shorter than the mean free time (or the interval between collisions), and both these times, in turn, are significantly less than the macroscopic relaxation time in the whole volume of the system. The whole theory of non-equilibrium processes in gases is based on the existence of these different time scales.

We can distinguish three stages of the non-equilibrium process in the gas.

In the initial stage, if we are interested in the state of the system for time scales smaller than the collision time, the state of the gas depends strongly on the initial state, and to describe the non-equilibrium process it is necessary to define a large number of distribution functions or trajectories in the classical case.

For the time scales greater than the collision time there is a kinetic stage when the initial state is negligible and the system 'forgets' about it. In this case, a further reduction in its description is possible. To describe the state of the system, it is sufficient to have the one-particle distribution function, and other higher-order distribution function depend on time only through the one-particle distribution function. The kinetic stage is sufficient to use the Boltzmann kinetic equation for the one-particle distribution function.

For the time scales exceeding the mean free time, there comes a stage of the hydrodynamic non-equilibrium process, where a further reduction in the description of the system is possible. Now it suffices to set the temperature, chemical potential and the mass flow rate, or the average energy density of particles and momentum. At the hydrodynamic stage it is possible to construct the hydrodynamic equations, i.e. equations of heat conduction, diffusion and transfer equations of momentum.

For a fluid the situation is much more complicated than in the case of a gas or a solid. The point is that the concept of binary collisions loses its meaning together with the concept of collision time and mean free path time. These quantities are replaced by the correlation times between the dynamic variables. The non-equilibrium processes in fluids can have an initial stage required to describe a large number of distribution functions, and a hydrodynamic stage, described by the set of hydrodynamic parameters: temperature, chemical potential and mass velocity. The latter begins when the characteristic times of changes in hydrodynamic parameters are considerably greater than the correlation time between the dynamic variables.

Thermodynamics of irreversible processes is becoming more widely used in various fields of the physics of gases and solids. A large area of its application are the theoretical problems of radiation physics of solids, since in the conditions of irradiation with atomic particles the crystal is a clear example of a strongly non-equilibrium system. The subject of investigation is the elementary act of interaction of fast particles with the crystal, the theory of primary defects, including in a cascade of processes, their further evolution, and the movement of high-energy particles in the channeling mode. We have attempted to focus the reader's attention to new research in radiation physics, to identify the main trends of its development and nature of basic research in this area.

A large class of irreversible processes can be interpreted as the system response to an external mechanical disturbance. However, this does not exhaust the list of all irreversible processes, since there are physical phenomena caused by thermal perturbations, primarily temperature gradients. It should be noted that the separation of the perturbations into mechanical and thermal makes sense only in the linear approximation, since the higher approximations of mechanical perturbations create inhomogeneities in the distribution of particles and momentum and this, in turn, gives rise to thermal perturbations. All types of perturbations are considered by non-equilibrium thermo-dynamics, which is based on the method of statistical ensembles for non-equilibrium systems. The method of statistical Gibbs ensembles for non-equilibrium systems was developed in two directions, namely, as a method of the non-equilibrium statistical operator, based on the construction of local integrals of motion [4], and as a method, based on taking into account the effect of the thermostat through the non-potential forces [5]. Both approaches lead to virtually identical results, but the approach proposed in [4] is more convenient for application to specific problems.

Of course, problems of radiation physics of solids require, in addition to the method, the corresponding model representations. The model of the theory of displacement cascades and the theory of positronium atom beams are discussed in Chapter 7. As for the models used in the theory of the channeling effect, they are more complex and require detailed discussion. Therefore, they will be dealt with in a separate chapter.

Research interests of the author include mainly the study of non-equilibrium processes observed during the movement of atomic particles in crystals. Such a system (crystal and fast particles moving in it) is a typical non-equilibrium closed system. Another type of system – open non-equilibrium systems – are more common in problems of chemical kinetics. If the system is closed, then the second law of thermodynamics is satisfied, whereas an open system can be characterized by a decrease in entropy and, consequently, by the growth of ordering (self-organization). Only closed systems will be considered in the future. Considered are the stochastic theory of motion of high energy particles in planar channels (Chapter 2), the method of the non-equilibrium statistical operator and its application to problems of non-equilibrium thermodynamics of channeled particles (Chapters 5, 6), the non-equilibrium thermodynamics of chaotic particle beams and the relaxation kinetics of the cascade particles (Chapter 7). Moreover, Chapters 2–4 are devoted to problems of non-equilibrium statistical mechanics, and Chapters 5–7 deal with non-equilibrium statistical thermodynamics. The book primarily presents new results obtained in recent years, with only the minimum data presented in earlier studies. All the latest studies are referenced. At the same time, the system of units in which Planck's constant and Boltzmann's constants are equal to unity, $\hbar = k_B = 1$, is used.

1.3. Channeling effect of charged particles

1.3.1. A brief overview

Passage of a beam of charged particles through matter is accompanied by a range of physical phenomena caused by particle collisions with atoms of matter: inhibition of beam particles in the material, changing the direction of the velocity of their movement, the yield of nuclear reactions and X-ray radiation, changes in the structure of matter, the formation of new particles, etc. For an isotropic distribution of atoms of matter, the probability of these events taking place depends only on the number density of atoms, their properties and structure and does not depend on the direction of motion of particles in matter.

However, the strictly ordered arrangement of atoms can lead to qualitatively new features of the interaction of particles with single crystals. Indeed, in a single crystal in the direction of the main crystallographic axes there are tightly packed chains of atoms and the distances between the chains is maximum. Located in a strict order, the atoms of these chains create walls of the channels that are open to the motion of particles. The particles whose velocities are directed along the channel axes are less likely to disperse than the particles that move in an arbitrary direction. There are many such channels in the crystal, but the most open to the motion of the particles are channels that are located between the most densely packed rows of atoms. For the same particles emitted from the crystal lattice one can observe the opposite pattern: the particles whose velocities are directed along the crystallographic axes with low indices will be more likely to be scattered by the atoms of the crystal than the particles with the arbitrary orientation of the velocity direction. These visual representations make it possible to understand the nature of the phenomenon of channeling and shadow effects, which are collectively called directional effects in crystals.

The possibility of a significant impact on the character of periodicity of the structure of particles in the crystal was suggested for the first time by Stark (1912). But then the level of development of accelerator technology was insufficient to generate the necessary high-energy particle beams so that these effects could not be observed experimentally. Only in the course of experiments on single-crystal sputtering targets (1954) was it possible to detect the anisotropy of the ion sputtering yield associated with the directions of the main crystallographic axes. In further experiments [6] carried out to introduce heavy ions in crystalline targets of aluminum and tungsten it was shown that an unexpectedly large number of ions penetrate the anomalously great depths.

Some time later, the direct simulation of particle trajectories in the crystal showed [7] that the anomalously large penetration depth is typical only of those particles whose trajectories are in the direction of the most densely packed rows of atoms between which there are open channels. Particles trapped in these channels at a relatively small angle to the direction of the axes of the channels are held in them by moving large numbers of collisions with the atoms of these channels. Robinson and Owen [7] called this phenomenon channeling. The existence of channeling was subsequently demonstrated experimentally by many researchers. Thus, in the experiment [8] in which gold crystals with a thickness of 3000 Å were bombarded by a beam of protons with energies of 75 keV, it was found that during the rotation of the crystal the flux of ions passing through it increased approximately tenfold, if the incident ion beam is parallel to the crystallographic axis (110).

Basic theoretical concepts of channeling were developed in [9, 10]. In these studies, channeling was considered as a phenomenon caused by a large number of correlated collisions of beam particles with tightly packed rows of atoms of the crystal. In each row, the particle is deflected from the direction of its movement by a small angle. The most consistent and complete theory of channeling was proposed by Lindhard [10]. His theory is based on the interaction of particles with chains and planes of atoms by means of continuous potentials obtained by averaging the potentials of the static atomic chain (or plane) along the chain (or plane). At the same time, the arrangement of atoms in a crystal was considered perfectly periodic.

Channeling is due to the fundamental difference between the interaction of fast charged particles with the crystal atoms as they move at small angles to crystallographic planes or to low-index crystallographic planes or axes of their interaction with non-oriented movement. Features of the interaction of channeled particles with the atoms of the crystal assuming the applicability of classical mechanics position and momentum approximation are as follows:

– In motion of particles in a crystal at angles less than critical angle ψ_{cr}, relative to close-packed crystallographic planes (axes), the crystal as a whole forms a stable trajectory of a particle through the averaged potentials of atomic planes (chains).

– Parameters of the pair interactions of the channeled particles with the atoms of the crystal and the characteristics of all the secondary processes induced by the channeling, are completely determined by the trajectory.

The first assertion means that the interaction of channeled particles with the crystal has a collective character, and the second – that their collisions with the atoms are strongly correlated, since the impact parameters of collisions are uniquely defined by the particle trajectory. Moreover, these parameters vary slightly from collision to collision. For this reason, the interaction of the particles with the atoms of a crystal having the above characteristics can be considered correlated.

In theoretical studies of non-equilibrium processes accompanying the passage of the channeled particles through a crystal, we would like to highlight two distinct areas. The first of these [10] is based on the diffusion model in which multiple scattering leads to a slow diffusion of particles in the space of transverse energies. In this model, using the diffusion stochastic equations, attention is given to the evolution of spatial (angular) and energy distribution of channeled particles, including the time evolution of the particle flux leaving the channeling regime. The main feature of the equations [10] is their nonlocality, in particular, the distribution function does not depend on the transverse coordinates of the particles.

At the same time, a more complete analysis of the results of experiments on the dechanneling theory requires the development of local non-equilibrium processes accompanying the motion of fast particles. Development of the local theory was studied in the second direction [11,12]. In these studies attention was focused on the construction methods of the modern theory of irreversible processes, the local kinetic equation for the distribution function of channeled particles, as well as the calculation of the kinetic coefficients.

Quantum-mechanical treatment is useful not only in studies of channeling in thin crystals, but also in thick crystals to determine the subtle effects, especially when studying the scattering of light particles (electrons, positrons). Application of the quantum-mechanical density matrix for the study of channeling with multiple scattering taken into account is the subject of papers [13,14]. It was found that at some depth L_{coh}, called the coherence length, the stage of statistical equilibrium, in which the distribution of the channeled particle velocity is symmetric and quasi-stationary, begins.

This led to a significant simplification of further theoretical analysis. However, the calculations were completed only for the planar potentials of the step type. In [15–17], another approach based on the application of the methods of the statistical operator and Green functions, is proposed. With this approach, it was possible to obtain analytical expressions for the rate of dechanneling and the flux of particles at the stage of approaching the quasi-equilibrium.

1.3.2. Some assumptions of the classical theory of channeling

Lindhard's theory [10] reviews directional effects, in particular, the channeling effect, based on the laws of classical mechanics. In the experimentally accessible range of velocities the condition of the classical approach, as shown by estimates, is efficiently satisfied for protons and heavier atomic particles moving in a solid.

The periodicity of the crystal lattice arrangement of atoms significantly affects the motion of charged particles in a single crystal. When there is a match within a certain critical angle [10] between the direction of the beam and the crystal plane (axis), the correlation in the scattering of particles by atoms of the lattice, fixed in equilibrium positions, becomes very strong.

Under these conditions there are only correlated collisions with atoms, forming a crystallographic plane (series) and, consequently, the particles are trapped in the plane (axial) channel. These particles are indispersible at large angles, as in the channeling regime the processes that need to implement a collision with small impact parameters are suppressed. In addition, under channeling conditions the angular distribution of the particles [10] and the energy loss of the beam change.

To examine the effect of correlated deviations at small angles on the motion of a fast particle, the authors of [10] introduced an approximation of a continuous atomic chain, according to which the deviation of the particle trajectory is due to many successive collisions with atoms in the chain. The total effect of a sequence of classical collisions is described by a continuous potential at distance ρ in a plane perpendicular to the chain axis. This potential is found by averaging $V(r)$ – the potential of the interaction of atomic particle, passing along the axis z, with the lattice atom.

As a result, for the axial channel, we obtain

$$U_a(\rho) = \frac{1}{d}\int_{-\infty}^{\infty} dz\, V\left(\sqrt{\rho^2 + z^2}\right),\tag{1.4}$$

where $r = (\rho, z)$, d is the distance between the nearest atoms in the chain.

Similarly, we consider the motion of a particle almost parallel to the crystallographic plane. In this case, the potential is determined by the uniformly filled plane. Let the positively charged particles fly into the crystal at an angle ψ_0 relative to the crystallographic plane forming the channel wall. The particles lying on the segment length $(mv\psi_0)^{-1}$ then coherently interact with all atoms of the plane. (Here m and v is the mass and velocity of the fast particles, respectively).

If this length is large compared with the interatomic distance, then the arrangement of atoms in the scattering plane is unimportant. Therefore, in the first approximation it is sufficient to consider the total potential of the plane, averaged over the arrangement of the atoms. After averaging, the continuous potential of the atomic plane as a function of the distance of the particle from the plane in the x-direction is expressed as follows

$$U_{pl}(x) = n_c d 2\pi \int_0^{\infty} d\eta\, \eta V\left(\sqrt{x^2 + \eta^2}\right),\tag{1.5}$$

where $r = (x, \eta)$, n_c is the density of atoms in the lattice.

Using the Thomas–Fermi potential as the atom–atom potential

$$V_{TF}(r) = \frac{Z_1 Z_2 e^2}{r}\chi\left(\frac{r}{a_{TF}}\right),\tag{1.6}$$

we calculate the continuous potential of the atomic chain and the planes. In the formula (1.6) χ is the Fermi screening function [18];

$a_{TF} = 0.89 a_B \left(Z_1^{1/3} + Z_2^{1/3} \right)^{-2/3}$ is the radius of Thomas–Fermi screening, a_B is the Bohr radius; Z_1 and Z_2 are charges the fast particle and the atom lattice. Substituting the Thomas–Fermi potential (1.6) into (1.4) and (1.5), we obtain the so-called standard Lindhard potential (axial channel)

$$U_a(\rho) = \frac{2Z_1 Z_2 e^2}{d} \ln \left[1 + 3\left(\frac{a_{TF}}{\rho} \right)^2 \right] \qquad (1.7)$$

and the continuous planar potential

$$U_{pl}(x) = n_c d 2\pi Z_1 Z_2 e^2 \left[\left(x^2 + 3a_{TF}^2 \right)^{1/2} - x \right]. \qquad (1.8)$$

According to estimates [10], the continuous potential approximation is valid in the case when the particles strike the crystal at sufficiently small angles ψ_0 to the atomic chains or atomic planes. At particle energies $E_0 > E'$, where $E' = 2Z_1 Z_2 e^2 d / a_{TF}^2$, the condition for ψ_0 can be written as

$$\psi_0 < \psi_{c1} = \left(\frac{1}{dE_0} 2Z_1 Z_2 e^2 \right)^{1/2}. \qquad (1.9)$$

If $E_0 < E'$, the approximation of the continuous potential can be used for angles

$$\psi_0 < \psi_{c2} = \left(\sqrt{3} a_{TF} \psi_{c1} \frac{1}{\sqrt{2}d} \right)^{1/2}.$$

In Lindhard's theory ψ_{c1} is the angle between the initial particle velocity v_0 and the atomic chain, in which the interaction of particles with the chain is no longer correlated. Of course, the more precise definition of the critical angle requires taking into account the impact on the particle of not a single atomic chain but the channel as a whole. For this purpose, we can use a simple model which nevertheless takes into account the finiteness of the transverse dimension of the channel. Namely, we consider a channel which is formed by two parallel atomic chains lying in one plane. As shown by calculations, the exact equations for the critical angle in this case are very complex and can solved only by the numerical integration methods.

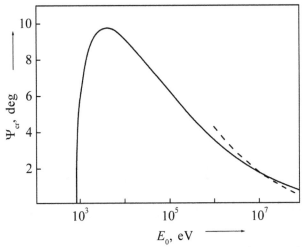

Fig. 1.1. Dependence of the critical channeling angle on particle energy.

Numerical calculation of the critical angle when a copper atom moves between two chains [100], also consisting of copper atoms, was performed in [19]. The corresponding graph is shown in Fig. 1.1, where the dotted line shows the dependence of the critical Lindhard angle (1.9) on particle energy. As can be seen from the figure, the allowance for the finite transverse dimensions of the channel really leads to the formation of a maximum of the function with the critical angle converting to zero at low energies. Such behavior differs fundamentally from the behavior ψ_{c2}. As for the high-energy range, then, as shown in the figure, the functions in this range differ only slightly.

It should be added to this that the existence of a finite critical angle is related not only to the loss of correlation of particle collisions with the walls of the channel, but also to the emergence of a parametric instability of motion of channeled particles [19].

2

Channeling of charged high-energy particles as a stochastic process

2.1. Markov process with a normal (Gaussian) distribution

In a large class of systems situations can occur in which the laws of motion of the particles deviate from the laws of mechanics. In this case, in the equations of motion of the system we introduce some elements that have a probabilistic nature. Random, or stochastic, is a process in which the coordinate of the process $\lambda(t)$ is not fully defined depending on the variable t. The random variable $\lambda(t)$ can have a certain value with some probability and is determined if a set of possible values and the probability of each of these values are given.

In order to study the probability distribution, the following conditional probability is

$$P_2 \left(\lambda_1, t_1 \mid \lambda_2, t_2 \right) d\lambda_2,$$

which represents the probability of finding $\lambda_2(t)$ in the interval $(\lambda_2, \lambda_2 + d\lambda_2)$ at time t_2, provided that $\lambda_1(t) = \lambda_1$ for all $t_1 \leq t_2$. We assume that this condition characterizes the process for which the background observation that preceded the last observation time, i.e. time t_1, has no effect on the probability distribution. Such a class of stochastic processes forms the so-called Markov processes.

A classic example of a Markov process is the Brownian motion. Brownian motion refers to movement of a body caused by environmental shocks of particles in thermal motion. The magnitude and sign of displacement of a Brownian particle between two collisions are independent of previous collisions and, in this sense, the random walk should be considered as a Markov process. Typically, the study of Brownian motion investigates the motion of macroscopic bodies. However, the results obtained in this study have a high degree of generality and are usually applied to individual atoms and molecules.

The Gaussian probability distribution is used most often as the probability density of a Markov process. Many two-time correlation functions in the theory of Markov processes have the same form. For example, in the presence of independent increments of the momenta of Brownian particles, the corresponding force–force correlation function can be written in Gaussian form

$$\left\langle \tilde{f}_B(t)\tilde{f}_B(t') \right\rangle_G = \frac{1}{2\tau_k}\gamma \exp\left\{-\frac{|t-t'|}{\tau_k}\right\}. \qquad (2.1)$$

Here, $\tilde{f}_B(t)$ is random force; τ_k is the time correlation of forces; γ is a positive definite quantity. Averaging $\langle...\rangle_G$ in (2.1) is satisfied with a normal (Gaussian) distribution

$$n_G(\lambda) = \frac{1}{\sqrt{2\pi}\sigma_\lambda}\exp\left\{-\frac{(\lambda-\overline{\lambda})^2}{2\sigma_\lambda^2}\right\}, \qquad (2.2)$$

which includes two parameters: $\overline{\lambda}$ – mathematical expectation and σ_λ^2 – variance. In general, both parameters may vary from time to time, but in this book only stationary Markov processes are considered.

If the time interval $|t - t'|$ is much larger than the correlation time τ_k, then, as follows from (2.1), forces $\tilde{f}_B(t)$ and $\tilde{f}_B(t')$ are uncorrelated. Indeed, if $|t - t'| \gg \tau_k$ the right side of (2.1) is close to zero, while for $|t - t'| \ll \tau_k$ it takes the value $\gamma/2\tau_k$. So in the limit $\tau_k \to 0$ we have

$$\left\langle \tilde{f}_B(t)\tilde{f}_B(t') \right\rangle_G = \gamma\delta(t-t'), \qquad (2.3)$$

where $\delta(t)$ is the delta-function. In other words, in the limit $\tau_k \to 0$ we obtain a stochastic process with the zero mean value, which is similar to 'white noise' [20]. We note that because of stationarity and homogeneity of the equilibrium distribution, the correlation function (2.3) depends only on the difference $t - t'$.

2.2. Transport coefficients of channeled particles in electron scattering (non-local theory)

2.2.1. Fokker–Planck equation with fluctuations
The diffusion function and the friction coefficient

This section focuses on the analysis of the properties of random forces caused by the interaction of fast particles with the electron gas. It is assumed that the electrons are distributed uniformly throughout the volume of solids. Homogeneity determines the non-locality of the following proposed theory.

With the passage of fast particles through an electron gas the interaction with electrons appears as the effect on the particle of a random polarization field $\mathbf{E}^p(t)$ and a random electric field $\mathbf{E}^{fl}(t)$, due to fluctuations of electron density. As for the source of the field $\mathbf{E}^p(t)$, the appearance of polarization in the many-electron system is associated with the following phenomenon. If a screening cloud forms in the vicinity of a positively charged particle Z_1 at rest in an electric gas, then in the uniform motion of the charge the center of gravity of a negatively charged cloud lags behind the particle, and the length of the polarization charge distribution is estimated as $(m_e v)^{-1}$, where m_e is the electron mass. The so-called wake polarization of the medium results in the formation of an electric field $\mathbf{E}^p(t)$, retarding the motion of channeled particles (CP).

The force acting on the particle in the total field $\mathbf{E}^p + \mathbf{E}^{fl}$, with the accuracy up to second order in the fluctuating component can be written as

$$\tilde{f}_x^{(e)}(t) = Z_1 e\left[E_x^p(t) + E_x^{fl}(t)\right] +$$
$$+ (Z_1 e)^2 \int_{t_0}^{t} dt' \int_{t_0}^{t'} dt'' E_x^{fl}(t'') \frac{\partial}{\partial x}\left[E_x^p(t) + E_x^{fl}(t)\right].$$
(2.4)

Equation (2.4) corresponds to the geometry of the particle in a planar channel, and the value of t_0 in (2.4) is considered as the origin of time. However, we must bear in mind that in the future we will deal only with stationary random processes, for which the probability distributions are invariant under the shift of the origin of time.

Polarization force $\tilde{f}_x^p(t) = Z_1 e E_x^p(t)$ is expressed through the imaginary part of the Fourier component of the longitudinal dielectric permittivity $\varepsilon_L(k,\omega)$ [21]

$$E_x^p(t) = \frac{1}{2\pi^2} Z_1 e \int d\mathbf{k} \frac{k_x}{k^2} \operatorname{Im} \varepsilon_L^{-1}(\mathbf{k}, \mathbf{k} \cdot \mathbf{v}(t)). \qquad (2.5)$$

As for the fluctuation force $\tilde{f}_x^{fl}(t) = Z_1 e E_x^{fl}(t)$, it is convenient to express this force using the spectral representation of the correlation function for the field strength. Namely, using the spectral density

$$\left\langle E_x^{fl} E_x^{fl*} \right\rangle_{\mathbf{k},\omega} = 24\pi^3 \frac{e^2 n_e}{k^3 \left|\varepsilon_L\right|^2 v_F} \left(1 - \frac{\omega^2}{k^2 v_F^2}\right) \left(\frac{k_x}{k}\right)^2 \theta\left(1 - \left|\frac{\omega}{k v_F}\right|\right), \qquad (2.6)$$

where n_e – electron density; v_F – Fermi velocity; $\theta(x)$ – theta function.

Given the existence of a random force (2.4), a stochastic equation of motion of CP in the cross direction relative to the plane of the channel can be written as

$$m\ddot{x}(t) = -\frac{1}{v}\dot{x}(t)\left(-\frac{dE}{dx}\right)_e - \nabla_x U_{pl}(x) + \tilde{f}_x^e(t), \qquad (2.7)$$

where $U_{pl}(x)$ – constant potential planar channel; $(-dE/dx)_e$ – electron energy loss per unit path length.

As in the case of polarization, in the case of the fluctuation force it is also not difficult to form a temporal correlation function of the force – a force that gives a measure of correlations between random effects at different times. These correlation functions include the finite correlation time, and this can be considered as a circumstance that prevents the use of the mathematical apparatus of Markov processes. However, in practice it is often necessary to deal with systems for which the external influence is not delta-correlated (2.3).

In fact, the scope of the theory of Markov processes is much broader [20]. In particular, the movement of high-energy particles, impacted by the random force (2.4), can be regarded as a Markov process, if $\omega < kv$. Taking into account (2.6), it is assumed that fluctuations of the spectrum do not include waves with phase velocities $|\omega/k| > v_F$, so that for the particles with velocities $v > v_F$ and $v \gg v_F$ the above condition is satisfied. In turn, this circumstance allows us to use the standard methods of the theory of Markov processes for constructing Fokker–Planck equations for the probability density $\mathcal{P}(x, vx, t)$ in the transverse space (x, v_x).

Thus, the stochastic equation (2.7) corresponds to the Fokker–Planck equation describing the time course of the Markov process

$$\frac{\partial}{\partial t} P(x, v_x, t) + v_x \frac{\partial}{\partial x} P +$$

$$+ \frac{\partial}{\partial v_x} \left[-\frac{\partial U_{pl}(x)}{\partial x} + \xi_x^P + \xi_x^{fl} \right] P - \frac{1}{2} \frac{\partial^2}{\partial v_x^2} \mathcal{D}_{xx} P = 0. \tag{2.8}$$

Here $\mathcal{D}_{xx} = \mathcal{D}_{xx}^P + \mathcal{D}_{xx}^{fl}$ is the diffusion function in space of the transverse velocity, which characterizes a stochastic process due to random forces $\tilde{f}_x^P$ and $\tilde{f}_x^{fl}$. This diffusion stochastic process can be regarded as a generalization of the Wiener process [20]. In addition, (2.8) includes $\xi_x = \xi_x^P + \xi_x^{fl}$ – the dynamic friction coefficient. Kinetic functions (coefficients) are

$$\mathcal{D}_{xx} = \mathcal{D}_{xx}^{fl} + \mathcal{D}_{xx}^P,$$

$$\mathcal{D}_{xx}^P = \left(v\tilde{\mu} \right)^{-2} \left(v_x^2 - v^2 \right) K_{xx},$$

$$\mathcal{D}_{xx}^{fl} = \frac{1}{\tilde{\mu}} G_{xx}, \tag{2.9}$$

$$\xi_x^P = \frac{1}{\tilde{\mu}} \left(\frac{v_x}{\langle v^2 \rangle} \right) K_{xx}, \qquad \xi_x^{fl} = \frac{1}{2} \left(\frac{\partial G_{xx}}{\partial v} \right)$$

and expressed in terms of the correlation function and power spectral density $\tilde{f}_x^P$ (2.6)

$$K_{xx} = \int_{-\infty}^{\infty} d\tau \left\langle \tilde{f}_x^P \tilde{f}_x^P (\tau) \right\rangle,$$

$$G_{xx} = \frac{(Z_1 e)^2}{(2\pi)^3} \int d\mathbf{k} \left\langle E_x^{fl} E_x^{fl*} \right\rangle_{k_s(kv)}. \tag{2.10}$$

In the formula (2.9) $\tilde{\mu}$ is the reduced mass of the channeled particle (CP). Note also that in what follows we consider only the normal stochastic processes and, in some cases, the index G in the notation for the average is omitted.

The Fokker–Planck equation (2.8) describes the superposition of two processes: diffusion and friction in the space of transverse velocity. If the CP distribution at the initial time has a maximum near the entry energy of the particles into the channel, under the influence of friction this peak will shift toward lower energy values (mean transverse energy). In addition, the maximum of the distribution is broadened considerably with time, so another physical quantity – the dispersion process – becomes essential for understanding the whole process.

The channeling of high-energy particles in a metal is often studied using a simplified model in which the valence electrons are considered as a degenerate Fermi gas with high density n_V at absolute zero temperature. In this model, particles with impact parameters $b \leq k_F^{-1}$ participate in the uncorrelated pair collisions with electrons, but in other cases the particles interact with the collective modes that are characteristic for a dense electron gas, i.e. with plasma oscillations. Here $k_F = m_e v_F$ is the Fermi momentum.

In order to express the kinetic functions (2.9) through the main parameters of the microscopic theory, we must first calculate the correlation function K_{xx} (2.10) in the case of high velocities $v > v_F$. Using the explicit form $\tilde{f}_x^p$ (2.5) and the longitudinal dielectric constant of a dense electron gas, written in the form [22], after several transformations we obtain

$$K_{xx} = \frac{4\pi Z_1^2 e^4}{v} n_e \ln\left(\frac{b_{max}}{b_{min}}\right). \qquad (2.11)$$

In (2.11), as already noted, as the maximum impact parameter can take the value of $b_{max} = k_F^{-1}$, and as the lowest value possible to take the impact parameter, known from the uncertainty relation $b_{min} = (m_e v)^{-1}$. Given the form of these parameters, as well as the type K_{xx} (2.11), G_{xx} (2.10) and (2.6) after substituting these quantities in (2.9) we obtain the final expression for the diagonal elements of the diffusion tensor and dynamic friction coefficients in the case of $v > v_F$. Namely,

$$D_{xx} = \frac{4\pi Z_1^2 e^4}{m_e^2 v} n_V \left\{ L_e + \left[\left(\frac{v_x}{v}\right)^2 - 1 \right] \ln\left(\frac{m_e v}{k_F}\right) \right\},$$

$$\xi_x^{fl} = \frac{4\pi Z_1^2 e^4}{m_e v^2} n_V L_e, \qquad L_e = \ln\left|\frac{2m_e v v_F}{\omega_p}\right|, \qquad (2.12$$

$$\xi_x^p = \frac{v_x}{v} \frac{4\pi Z_1^2 e^4}{m_e v^2} n_V \ln\left|\frac{m_e v}{k_F}\right|,$$

where ω_p is the frequency of plasma oscillations. As seen from (2.12), the friction coefficient corresponding to the first-order term in the Fokker–Planck equation (2.8) is proportional to the electron energy loss per unit length $(dE/dx)_V$.

2.2.2. Stochastic theory for the thermodynamic level of description

Relations describing the change of any macroscopic system should cover any thermodynamic aspect of the theory. We recall the well-known example. In Onsager' theory, the relaxation of macroscopic systems to equilibrium is due to the presence of thermodynamic forces, and fluctuations, such as random noise in the thermodynamic flows, are ultimately related to the thermodynamic quantities. It is therefore natural to include any aspect of the thermodynamic formalism based on the microscopic mechanism of the elementary process. So far we have considered the motion of fast particles from the standpoint of mechanics, but it would be better to go to the thermodynamic level of description. Such a transition can be at first carried out conveniently on the example of a simpler system than the fast particles. Namely, for example, the motion of Brownian particles. The theory of Brownian motion is the subject of extensive literature [20,23,24], and we refer the reader to it. Here, we note only that the model used in the theory of Brownian particles is similar to the model used in the theory of channeling where an atomic particle moves in a gas of light particles – in the electron gas.

It is easy to see how the momentum of a heavy Brownian particle changes with mass m_B, if we consider the effect on the particle of a systematic force, namely, the friction force and, in addition, a random force due to the shocks of the molecules from the environment. Given the symmetry, the shocks experienced by the Brownian particle in different directions can be considered equally probable. Therefore, the average value of the random force $\tilde{f}_B$ is equal to zero. The time correlation of random forces τ_k to a first approximation can be considered equal to the average time interval between two successive shocks.

If the concentration of Brownian particles is small and the gas concentration is large enough, for the time intervals $\Delta t \gg \tau_k$ the Brownian particle undergoes many collisions. Therefore, according to the limit theorem in the theory of stochastic processes [20], at times of the order Δt the value of $\tilde{f}_B$ can be regarded as a normal Markov process with the delta-shaped force–force correlation function. This physical situation greatly simplifies the expressions of the theory of

Brownian motion. The approximations introduced into the theory, as discussed below, are ultimately justified by the fact that we are interested in the macroscopic (and not microscopic) picture of the motion of heavy particles.

The main kinetic coefficient of the theory of Brownian motion is the friction coefficient ζ_B. In a situation which we would like to analyze, all the stochastic quantities are very weakly dependent on time and it is sufficient to study a stationary Gaussian Markov process [20]. In this case, non-equilibrium statistical mechanics gives the following expression for the friction coefficient [25]

$$\zeta_{\scriptscriptstyle{SB}} = \frac{1}{3m_{\scriptscriptstyle B}}\beta_1 \int_{-\infty}^{\infty} d\tau \langle \tilde{f}_{\scriptscriptstyle B}(0)\tilde{f}_{\scriptscriptstyle B}(\tau)\rangle_{\scriptscriptstyle G},\qquad(2.13)$$

where β_1 is the inverse temperature of the thermostat, i.e. gas of light particles.

Now imagine a completely different approach to the same problem, namely the approach based on the Liouville equation. To begin with, we write the total Hamiltonian of the system. It consists of medium Hamiltonian H_0, the kinetic energy of a Brownian particle T_k and the Hamiltonian of the interaction of the Brownian particle with the thermostat H_1

$$H = H_0 + T_k + H_1.$$

The form of the Hamiltonian determines the specific form of the Liouville operator L (1.3) and thus also the equation for the distribution $\rho(t) = \rho_l(t) + \Delta\rho(t)$

$$\left(\frac{\partial}{\partial t} + iL + \epsilon\right)\Delta\rho(t) = -\left(\frac{\partial}{\partial t} + iL\right)\rho_l(t).\qquad(2.14)$$

Here $\epsilon \to +0$ has the meaning of an infinitely small source which breaks the symmetry of (2.14) with respect to time reversal, and

$$\rho_l(t) = \exp\{-S(t)\} = \exp\{-\Phi_1(t) - \beta_1 H - \chi(t)p\}\qquad(2.15)$$

is the quasi-equilibrium distribution of particles. Note that the source is included in that term of equation (2.14) which describes the relaxation of the distribution $\rho(t)$ to the quasi-equilibrium with the average time $1/\epsilon$. As usual, in (2.15) $\Phi_1(t)$ is the Massieu–Planck function, χ is a parameter which involves the average momentum of a Brownian particle $\langle p\rangle^l = \mathrm{Sp}(\rho(t)p)$. Moreover, χ is found from the self-consistency condition, i.e. from the condition that the average momentum is equal to its quasi-equilibrium value.

In the classical case, the solution of (2.14) has the form

$$\Delta \rho(t) = i \int_{-\infty}^{t} d\tau \exp\{\epsilon(t-\tau)\} LS(\tau)\rho_{l}(\tau)\exp\{i(\tau-t)L\}. \quad (2.16)$$

Using the distribution (2.15) and expression (2.16), we find the equation for the rate of change of the momentum of a Brownian particle

$$\frac{d}{dt}\langle p \rangle^{t} = -\zeta_{B}\langle p \rangle^{t},$$

where the right-hand side describes the loss of momentum under the impact of shocks from the environment. The right-hand side of the equation is proportional to the friction coefficient, which as a result of simple transformations can be written in the form

$$\zeta_{B} = \frac{1}{m_{B}^{2}\langle v_{B}^{2}\rangle} \int_{-\infty}^{0} d\tau \exp(\epsilon\tau)\langle \delta f_{B} \exp(i\tau L)\delta f_{B}\rangle_{l},$$

$$\langle A \rangle_{l} = \mathrm{Sp}(\rho_{l}A), \quad (2.17)$$

where v_{B} is the velocity of a Brownian particle.

Equation (2.17) relates the friction coefficient with the correlation function of the fluctuation components of the forces

$$\delta f_{B} = f_{B} - \langle f_{B}\rangle_{l}.$$

Comparing the friction coefficient (2.13) given by non-equilibrium statistical mechanics, with the coefficient (2.17) obtained on the basis of non-equilibrium thermodynamics, shows that it is sufficient to perform two transformations for transition to the thermodynamic level. First, the random force $\tilde{f}_{B}(t)$ is replaced by the fluctuating part δf_{B}, secondly, the average of the stationary Gaussian distribution n_{G} (2.2) is replaced by averaging with quasi-equilibrium distribution ρ_{l} (2.15). As for replacing $\tilde{f}_{B}(t)$ by δf_{B}, it should be borne in mind that the fluctuations, which are spontaneous deviations from the mean, can, just as $\tilde{f}_{B}(t)$, have no exact values at a fixed time. In this sense, the indicated change is fully justified. And more. In the Brownian motion, the subsystem of heavy particles is close to thermodynamic equilibrium with a thermostat, so we can assume that

$$m_B \left\langle v_B^2 \right\rangle = 3 / \beta_1.$$

The last relation should be borne in mind when comparing (2.13) with (2.17).

Given the above considerations, it is easy to transfer to the so-called mechanical variant of non-equilibrium thermodynamics [23]. This variant can be applied to the theory of channeling, writing the polarization friction coefficient (2.9) in a form close to (2.17). We have

$$\xi_x^p = \frac{v_x}{m_e \left\langle v^2 \right\rangle} \int_{-\infty}^{0} d\tau \exp(\epsilon\tau) \left\langle \delta f_x^p \exp(i\tau L) \delta f_x^p \right\rangle_l. \qquad (2.18)$$

The expression of the friction coefficient via the correlation function of forces acting on the particle was first obtained in [26]. However, the full force figures in the formula [26] instead of fluctuating components of the forces δf_x^p, included in (2.18). Since in the case of full forces the integral with infinite limits is zero, in [26] integration by $d\tau$ was stopped at some time corresponding to the correlation integral reaching the values of the 'plateau'. This, in turn, gave rise to the plateau problem, which was resolved much later [27]. In the present formulation of ξ_x^p (2.18) such a problem is absent.

2.3. Kinetic features of channeled particles in the presence of thermal lattice vibrations (local theory)

2.3.1. Local matrix of random effects

In this section, an example of fast particles, interacting with the thermal vibrations of lattice atoms, is used to show that not only the energy loss of particles but also the diffusion function in the space of transverse energies can be represented in the form of an integral transform of the local matrix of random effects. This representation can be viewed as a traditional development approach of the theory of motion of atomic particles in the environment. Since the arrangement in space of the scattering centers, i.e. lattice atoms, is heterogeneous, the theory of channeling is formulated as a local theory.

Restricting ourselves to the analysis of the motion of fast particles between the two crystallographic planes (planar channeling), we assume that the x-axis is perpendicular to the plane of the channel (y, z), and the z axis is directed along the particle beam. The continuous potential is then a function of the transverse (with respect to the channel wall) coordinate $U_{pl}(x)$. The explicit form of $U_{pl}(x)$ is presented below.

The particle density at a given point varies with time. If we consider the spatial and temporal correlations, the density at point **r** at time t affects the value of the density at point **r'** at time $t + t'$. Nevertheless, in special cases, it is not justified to take into account two types of correlations. For example, in motion of fast particles in the electron gas the dominant role is played by correlations with respect to time. This is well explained when identical MeV particles are scattered by electrons and the velocity of particles is close to the Fermi velocity.

However, the physical situation is quite different in the case of scattering of MeV particles by thermal vibrations of lattice atoms. In this case, the velocity of thermal motion of lattice atoms is several orders of magnitude smaller than the velocity of the particles, so there is every reason to ignore the motion of atoms during the passage of particles through the crystal. In other words, it is possible to consider the motion of particles in a static lattice with 'frozen' atoms scatter, taking into account only one kind of correlation, namely, the spatial correlation.

In accordance with (2.2) and (2.3), we introduce the correlation function of random forces $\tilde{\mathbf{f}}(\mathbf{r})$, depending on the spatial variables,

$$\Psi_{\alpha\beta}\left(\mathbf{q},v_0 t;\mathbf{q}',v_0 t'\right) =$$
$$= \sum_{\nu\mu}\left\langle \tilde{f}_{\alpha}\left(\mathbf{q}-\mathbf{q}_c^{\nu},v_0 t - z_c^{\nu}\right)\tilde{f}_{\beta}\left(\mathbf{q}'-\mathbf{q}_c^{\mu},v_0 t'-z_c^{\mu}\right)\right\rangle. \qquad (2.19)$$

Here $\mathbf{r} = (\mathbf{q}, v_0 t)$, $\mathbf{q} = (x, y)$, $v_0 t$ is the longitudinal coordinate of the particle; $\mathbf{r}_c^{\nu} = \left(\mathbf{q}_c^{\nu}, z_c^{\nu}\right)$, $\mathbf{q}_c^{\nu}$ is the coordinate of the ν-th lattice atom in a transverse plane relative to the axis z, z_c^{ν} is the longitudinal coordinate of the atom, the indices α and β take the values of x, y, z, v_0 is the initial velocity of the fast particle.

In the theory of relaxation processes [28] the effect of the lattice with a 'frozen' scatter of the atoms can be described by two variants: either with a local stochastic force $\tilde{\mathbf{f}}(\mathbf{r})$, or by its classical analog $\delta \mathbf{f}(\mathbf{r})$. Here

$$\delta \mathbf{f}(\mathbf{r}) = \mathbf{f}(\mathbf{r}) - \left\langle \mathbf{f}(\mathbf{r})\right\rangle_G$$

is the fluctuation component of the force, and averaging $\left\langle \ldots \right\rangle_G$ is performed by the normal (Gaussian) distribution of atoms. In accordance with this the force–force temporal correlation function (2.19) can be written as

$$\Psi_{\alpha\beta}\left(\mathbf{q},v_0t;\mathbf{q}',v_0t'\right)=$$

$$=\sum_{\nu\mu}\left\{\left\langle f_\alpha\left(\mathbf{q}-\mathbf{q}_c^\nu,v_0t-z_c^\nu\right)f_\beta\left(\mathbf{q}'-\mathbf{q}_c^\mu,v_0t'-z_c^\mu\right)\right\rangle_G-\right.$$ (2.20)

$$\left.-\left\langle f_\alpha\left(\mathbf{q}-\mathbf{q}_c^\nu,v_0t-z_c^\nu\right)\right\rangle_G\left\langle f_\beta\left(\mathbf{q}'-\mathbf{q}_c^\mu,v_0t'-z_c^\mu\right)\right\rangle_G\right\}.$$

If (2.20) is integrated over time, we get a local matrix of random effects of the environment on the particle which is used to establish the connection of the microscopic analytical form, i.e. correlation function (2.20), with macroscopic kinetic functions. Thus, integrating with respect to time, we obtain

$$L_{\alpha\beta}=\int_0^t dt'\,\psi_{\alpha\beta}\left(\mathbf{q},v_0t;\mathbf{q}',v_0t'\right).$$ (2.21)

Assuming that the crystal is close to statistical equilibrium, we introduce the equilibrium probability density distribution of atoms in the lattice in the transverse plane $w_\perp(\mathbf{q}_c^\nu)$ and density in the z-direction $w_z(z_c^\nu)$

If the potential of interaction of a fast particle with a lattice atom is denoted $V(r)$, then using (2.20) and (2.21), we can show that

$$L_{\alpha\beta}\left(\mathbf{r};t\right)=L_{\alpha\beta}^{(1)}\left(\mathbf{r};t\right)+L_{\alpha\beta}^{(2)}\left(\mathbf{r};t\right),$$

$$L_{\alpha\beta}^{(1)}\left(\mathbf{r};t\right)=\int d\mathbf{r}_c W_1\left(\mathbf{r}_c\right)\nabla_\alpha V\left(\left|\mathbf{r}-\mathbf{r}_c\right|\right)\int_0^t d\tau\nabla_\beta V\left(\left|\mathbf{r}-\mathbf{r}_c-\mathbf{v}_0\tau\right|\right),$$

$$L_{\alpha\beta}^{(2)}\left(\mathbf{r};t\right)=\int d\mathbf{r}_c^{(1)}\int d\mathbf{r}_c^{(2)}g\left(\mathbf{r}_c^{(1)},\mathbf{r}_c^{(2)}\right)\nabla_\alpha V\left(\left|\mathbf{r}-\mathbf{r}_c^{(1)}\right|\right)\times$$ (2.22)

$$\times\int_0^t d\tau\nabla_\beta V\left(\left|\mathbf{r}-\mathbf{r}_c^{(2)}-\mathbf{v}_0\tau\right|\right).$$

Equation (2.22) uses the following notation: $W_1(\mathbf{r}_c)=w_\perp(\mathbf{q}_c)\,w_z(v_0t_1)-$ the one-particle function of the normal distribution of atoms;

$$g\left(\mathbf{r}_c^{(1)},\mathbf{r}_c^{(2)}\right)=W_1\left(\mathbf{q}_c^{(1)},v_0t_1\right)W_1\left(\mathbf{q}_c^{(2)},v_0t_2\right)-$$

$$-\tilde{P}_\perp\left(\mathbf{q}_c^{(1)},\mathbf{q}_c^{(2)}\right)\tilde{P}_z\left(v_0t_1,v_0t_2\right)$$

is the pair correlation function, $\nabla_\alpha=\partial/\partial r_\alpha,\tilde{P}_z\left(v_0t_1,v_0t_2\right)$ is the two-particle function of the normal distribution of the lattice atoms on the longitudinal coordinates, $\tilde{P}_\perp\left(q_c^{(1)},q_c^{(2)}\right)$ is the same function in the transverse plane.

2.3.2. Diagonal element of the local matrix. Correlation of random effects

Strictly speaking, the mathematical apparatus of Markov processes applicable to systems for which the external action is a random delta-correlated process [20]. However, the area of application of this theory is much broader. In particular, it can be used for the analysis of a random process with the zero mean value when the correlation function of forces is exponential. The finite correlation time of random effects does not preclude the application of the formalism of Markov processes, but only under certain conditions.

To clarify these terms, we introduce two relaxation times. First, τ_k – correlation time of forces, which by the order of magnitude is equal to the duration of a collision. Secondly, τ_c – correlation time of the particle velocity, which on the basis of the order of magnitude can be estimated as the mean free time. The stochastic process of the effect of the environment on the particle can be considered Markovian for $\delta t \geq \Delta$, if at all values of $(x, \varepsilon_\perp)$ the value of Δ satisfies the following inequalities

$$\min \tau_c > \Delta \gg \tau_k. \qquad (2.23)$$

The left inequality in (2.23) implies that the momentum of the particles during the time Δ varies only slightly. The right inequality means that the increment of the momentum at the non-overlapping time intervals of duration Δ are independent. The theory of solids shows [29] that the relaxation time τ_k is the smallest, so that the condition $\tau_c \gg \tau_k$ is well satisfied. This makes it possible to find such Δ at which the condition (2.23) is satisfied as a whole. Only in the case where (2.23) is satisfied we can use the kinetic equations and, in particular, the Fokker–Planck equation, which is part of the mathematical apparatus of the theory of Markov processes.

We introduce an additional approximation. According to (2.23), the time interval δt determines the upper limit of integration over time. Because this interval is the largest, it can be put equal to infinity. On this basis, in all subsequent formulas, the matrix (2.22) is replaced by its limit value – a fixed matrix

$$L_{\alpha\beta}(\mathbf{r}) = \lim_{t\to\infty} L_{\alpha\beta}(\mathbf{r};t).$$

Now we transform the expression (2.22). We assume that the distribution of atoms in the lattice plane of the channel is uniform. We can then assume that the elementary acts of scattering are due only to transverse displacements of lattice atoms u_x and in the matrix (2.22) its diagonal matrix element $\alpha = \beta = x$ is the main element. Given that $W_1(r_c) = n_c 2l w(x_c)$, where $w(x_c)$ is the one-particle distribution of the atoms in the transverse coordinate and n_c is the density of atoms in the lattice, the diagonal element of the matrix effects can be written as

$$L_{xx}^{(1)}(\mathbf{r}) = n_c 2l \int d\mathbf{r}_c w(x_c) \nabla_x V(|\mathbf{r} - \mathbf{r}_c|) \int d\tau \nabla_x V(|\mathbf{r} - \mathbf{r}_c - \mathbf{v}_0 \tau|). \quad (2.24)$$

In the case of the crystal structure with constant interplane distance $2l$ the one-particle distribution $w(x_c)$ is equal to $\sum_n \varphi(x_c - n2l)$ and because the averaging in (2.20) holds for a Gaussian ensemble, we have

$$\varphi(x) = (2\pi\sigma_\theta^2)^{-1/2} \exp\{-(x^2 / 2\sigma_\theta^2)\},$$

where σ_θ is the standard deviation of an atom in the direction perpendicular to the channel plane. Finally, the periodic function $w(x_c)$ has the form of a Fourier series

$$2l w(x_c) = \sum_n G_n \exp(ip_n x_c). \quad (2.25)$$

Here $p_n = \pi n / l$, n is the integer; $G_n = \exp\{-(p_n^2 \sigma_\theta^2 / 2)\}$.

As the interaction potential of the fast particle with the atom we select Moliere's approximation of the Thomas–Fermi potential

$$V_M(r) = V_c(r) \sum_i \alpha_i \exp(-\varkappa_i r), \quad V_c(r) = Z_1 Z_2 e^2 / r,$$

where $\alpha_1 = 0$, 1; $\alpha_2 = 0.55$; $\alpha_3 = 0.35$; $\varkappa_i = \beta_i / a_{TF}$; $\beta_1 = 6.0$, $\beta_2 = 1.2$; $\beta_3 = 0.3$; a_{TF} is the radius of Thomas--Fermi screening; Z_2 is the atomic number of the lattice atom; $i = 1, 2, 3$.

Substituting $W(x_c)$ and $V_M(r)$ into (2.24), we obtain an expression for $L_{xx}^{(1)}$. Unfortunately, this expression has a complicated analytical form. Even more complex is, as shown by calculation, the correlation correction $L_{xx}^{(2)}$. In order not to load the text with too long expressions, we restrict ourselves in this case to the representation of the approximation of the diagonal element of the matrix L_{xx}, as well as its graphic representation. To this end, we first introduce the dimensionless matrix element

$$\eta_{xx}(\tilde{x}) = \eta_1(\tilde{x}) - \eta_2(\tilde{x}),$$

$$\eta_1(\tilde{x}) = (Jm)^{-1} L_{xx}^{(1)}(\tilde{x}), \quad \eta_2(\tilde{x}) = -(Jm)^{-1} L_{xx}^{(2)}(\tilde{x}),$$

$$J = 8\pi \left(Z_1 Z_2 e^2 \right)^2 \frac{1}{mv_0} n_c.$$

The graph of the function $\eta(\tilde{x})$, where $\tilde{x} = x/l$, is shown in Fig. 2.1; the calculation were carried out for α-particles with the energy of 1.5 MeV, moving in a planar channel (100) of germanium, at temperatures $T = 300$, 500 and 700 K. The function $\eta(\tilde{x})$ is symmetric about the vertical axis and has the form of a bell-shaped (Gaussian) curve; the figure shows just the right-hand part. The full graph at $T = 300$ K is in Fig. 2.2.

The analysis showed that the matrix element $\eta(\tilde{x})$ can be approximated with reasonable accuracy by the difference of two Gaussian functions.

Namely,

$$\eta(\tilde{x}) = \left(\frac{300\text{K}}{2\pi T} \right)^{1/2} \left\{ \exp\left(-\frac{1}{2}\xi^2 (\tilde{x})^2 \right) - \frac{1}{5\xi x_0} \exp\left(-\frac{1}{2}\xi^2 (\tilde{x} - x_0)^2 \right) \right\}.$$

It was assumed that $1/\xi \approx 1 \text{Å}$ and $(\xi x_0)^2 = 1$. In general, the value of $1/\xi$ can be interpreted as the correlation length in the x-direction [23]. In the special case of a planar channel the role of the correlation parameter is played by the half-width of the

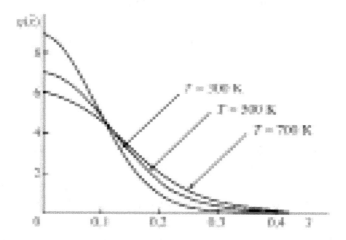

Fig. 2.1. Diagonal element of the local matrix of random effects as a function of the transverse coordinate.

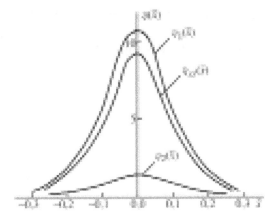

Fig. 2.2. The main term of the local matrix of random effects $\eta_1(\tilde{x})$ and the correlation correction $\eta_2(\tilde{x})$.

channel l, which determines the maximum deviation of the particle from the center of the channel.

We make two observations. First, the theory of correlation in space is far from complete, and in the literature considerations are generally limited to assessment of the correlation length. Imagine an assessment in this case $1/\xi \approx 1\text{Å}$. Second, the temporal variation of the matrix of random actions will affect the scattering of particles only if the time τ_ξ, during which the scattered particle is at the correlation length, comparable to the correlation of forces τ_k. However, the estimate for α-particles showed that at room temperature, the ratio τ_ξ/τ_k is negligible and, therefore, the contribution of temporal correlations is also negligible. In our opinion, arguments presented here confirm the validity of the approximation of static fluctuations and static correlation functions that we used in the study of scattering of fast particles by thermal vibrations of lattice atoms.

2.3.3. The system of stochastic equations and Fokker–Planck equation. Kinetic features of the local theory

The basic equations of the stochastic theory of channeling will be formulated. In the time hierarchy (2.23), the motion of fast particles in the crystal is a Markov process. If the variable is the transverse energy of the fast particle

$$\varepsilon_\perp = \frac{1}{2}m\left(\dot{x}\right)^2 + U_{pl}\left(x\right),$$

the equation for the joint probability density $P(x,\varepsilon_\perp,t)$ can be constructed on the basis of a system of stochastic equations corresponding to the two-dimensional Markov process $(x,\varepsilon_\perp)$

$(x,\varepsilon_\perp)$,

$$
\begin{cases}
\dot{x} = \left[\dfrac{2}{m}\left(\varepsilon_\perp - U_{pl}(x)\right)\right]^{1/2}, \\[4mm]
\dot{\varepsilon}_\perp \dfrac{1}{\left[\dfrac{2}{m}\left(\varepsilon_\perp - U_{pl}(x)\right)\right]^{1/2}} = -H\left(x,\sqrt{\dfrac{2}{m}\left(\varepsilon_\perp - U_{pl}(x)\right)}\right) + \tilde{f}(x).
\end{cases}
\tag{2.26}
$$

The second equation in (2.26) is in fact the Ito equation, written not in the traditional, but in a more convenient form, known as the non-linear Langevin equation. This equation includes the drift term $H\left(x,\sqrt{\dfrac{2}{m}\left(\varepsilon_\perp - U_{pl}(x)\right)}\right)$, which corresponds to the systematic part of the force associated with the external fields, and $\tilde{f}(x)$ is a random force due to the chaotic motion of atoms in the lattice.

The system of equations (2.26) corresponds to the statistically equivalent equation of motion for the probability $P(x,\varepsilon_\perp,t)$, i.e. the Fokker–Planck equation. The procedure for passing from (2.26) to the Fokker–Planck equation is well known [20], and there is no need to go into details of this transition. So imagine the final equation

$$
\begin{aligned}
\frac{\partial}{\partial x}P(x,\varepsilon_\perp,t) ={} & L(x)\frac{\partial^2}{\partial\varepsilon_\perp^2}\left\{\frac{1}{m}\left(\varepsilon_\perp - U_{pl}(x)\right)P(x,\varepsilon_\perp,t)\right\} - \\[2mm]
& -\frac{\partial}{\partial x}\left\{\left[\frac{2}{m}\left(\varepsilon_\perp - U_{pl}(x)\right)\right]^{1/2}P(x,\varepsilon_\perp,t)\right\} + \\[2mm]
& +\frac{\partial}{\partial\varepsilon_\perp}\left\{\left[\frac{2}{m}\left(\varepsilon_\perp - U_{pl}(x)\right)\right]^{1/2}\times \right. \\[2mm]
& \left. \times H\left(x,\sqrt{\frac{2}{m}\left(\varepsilon_\perp - U_{pl}(x)\right)}\right)P(x,\varepsilon_\perp,t)\right\} - \\[2mm]
& -\frac{1}{2}\frac{\partial}{\partial\varepsilon_\perp}\left\{\frac{1}{m}L(x)P(x,\varepsilon_\perp,t)\right\}.
\end{aligned}
\tag{2.27}
$$

Equation (2.27) includes the effects of diagonal element $L(x) = L^{(1)}(x)+L^{(2)}(x)$. The element brings in (2.27) the specificity of the motion of fast particles between the crystallographic planes in the presence of scattering by thermal vibrations of lattice atoms.

As the interaction potential of the fast particle with the atom we choose Moliere's approximation of the Thomas–Fermi potential V_M. Taking into account the explicit form of V_M and the distribution (2.25), as well as limiting the interaction of a fast particle with the nearest x-chain atoms in the lattice, the diagonal matrix element (2.24) is transformed to the form

$$L_{xx}^{(1)} \equiv L^{(1)}(x) = \frac{1}{v_0} 8\pi \left(Z_1 Z_2 e^2\right)^2 n_c \sum_n I_n G_n \cos\left(p_n x\right), \qquad (2.28)$$

where

$$I_n \equiv I(p_n) = \sum_{ij} \alpha_i \alpha_j \int \frac{dk\, k^3}{k^2 + \varkappa_i^2} \gamma_{nj}(k) \left\{ \tilde{\beta}_{nj}(k) \left[\tilde{\beta}_{nj}(k) + \gamma_{nj}(k) \right] \right\}^{-1},$$

$$\gamma_{nj}(k) = 2\left[k^2 + p_n^2 + \varkappa_j^2 \right],$$

$$\tilde{\beta}_{nj}(k) = \left[\left(\gamma_{nj}(k) \right)^2 - \left(4 p_n k \right)^2 \right]^{1/2}.$$

The expression for the correlation correction $L^{(2)}(x)$ is easy to write in the same form in which the primary member $L^{(1)}(x)$ (2.28) is written.

Since the interaction of fast particles with lattice atoms is assumed to be weak, transverse energy varies slowly with time. Given that the residence time of the process in the vicinity of a given value of x is inversely proportional to the speed, the conditional distribution at fixed transverse energy can be written as

$$P_c\left(x|\varepsilon_\perp\right) = \frac{1}{2\varkappa(\varepsilon_\perp)} \left[\frac{1}{m}\left(\varepsilon_\perp - U_{pl}(x)\right) \right]^{-1/2},$$

where

$$\varepsilon_\perp > U_{pl}(x), \quad \varkappa(\varepsilon_\perp) = \frac{1}{2} \int dx \left[\frac{1}{m}\left(\varepsilon_\perp - U_{pl}(x)\right) \right]^{-1/2}.$$

According to [6–14], the channeling effect occurs only in closed equilibrium systems in which there are no contributions to external factors such as external fields, the flow of final products, etc. Therefore, the drift term in the Fokker–Planck equation (2.27) is discarded. In this case, given that $\mathcal{P}(x,\varepsilon_\perp,t) = P(\varepsilon_\perp,t) P_c\left(x|\varepsilon_\perp\right)$, we represent the desired equation in the form

$$\frac{1}{J}\frac{\partial}{\partial t}P(\varepsilon_\perp,t) = -\frac{\partial}{\partial \varepsilon_\perp}\{a(\varepsilon_\perp)P(\varepsilon_\perp,t)\} +$$
$$+\frac{\partial^2}{\partial \varepsilon_\perp^2}\{b(\varepsilon_\perp)P(\varepsilon_\perp,t)\}\varepsilon_\perp^{cr},$$

(2.29)

where $a(\varepsilon_\perp)$ and $b(\varepsilon_\perp)$ are the dimensionless energy losses due to dynamic friction, and the dimensionless diffusion function describing the diffusion of particles in the space of transverse energy due to interaction with the lattice atoms. Given the explicit form of the diagonal elements of the random effects (2.22) and (2.24), kinetic functions $a(\varepsilon_\perp)$ and $b(\varepsilon_\perp)$ after some transformations can be represented in the final form

$$a(\varepsilon_\perp) = \frac{m^{1/2}}{2\varkappa(\varepsilon_\perp)}\sum_n \int dx \left[\varepsilon_\perp - U_{pl}(x)\right]^{-1/2} G_n K_n(x),$$
$$b(\varepsilon_\perp) = \frac{m^{1/2}}{\varkappa(\varepsilon_\perp)\varepsilon_\perp^{cr}}\sum_n \int dx \left[\varepsilon_\perp - U_{pl}(x)\right]^{1/2} G_n K_n(x).$$

(2.30)

Here $\varepsilon_\perp^{cr} = E_0\psi_{cr}^2 -$ the critical transverse energy. As for the planar channel potential $U_{pl}(x)$, it has the form of the potential obtained by averaging the Moliere approximation $V_M(x)$ on the distribution $w(x)$ [30]. In addition, (2.30) includes the factor

$$K_n(x) = I(p_n)\cos(p_n x),$$

describing the features of scattering in spatially inhomogeneous systems.

2.4. Dechanneling

2.4.1. The average lifetime of a dynamic system under random effects

As noted in Section 1.2, to describe the time evolution of a system consisting of N particles, we must know all $3N$ coordinates and $3N$ momenta of the particles as a function of time. Graphically, the time the development of such a system can be represented as a trajectory in $6N$-dimensional space (q, p), which is called the Γ-space. We assume that the point describing the state of an isolated system performs stable motion within a certain volume of the Γ-space.

A special feature of this class of problems which will be dealt with in the future is due to the presence of a closed boundary in the phase space. If the point representing the motion of the system reaches this limit, the system has a chance to disappear. In the task of channeling the emergence of such a boundary in the space of transverse variables is associated with the presence of the critical angle of channeling or critical transverse energy.

Consider a more realistic physical situation. Let the impact of the environment to be a random process due to which motion of the system ceases to be deterministic. Random shocks from the environment can lead to the fact that at some point in time the phase point reaches the boundary which we call the surface S. If the system reaches the surface S, then it is destroyed or absorbed, so that the problem arises of determining the average lifetime of a dynamic system subject to shocks.

If we assume that the random effects of environment on the channeled particles are delta-correlated, then when solving the problem of the lifetime it is natural to use the mathematical apparatus of Markov processes. Recall that delta correlation of pushes of a particle moving in an electron gas has already been discussed in Section 2.1.

As for the shocks caused by the thermal displacements of the atoms of the plane, the stochastic nature of the impacts of the atoms in a crystal undergoing thermal fluctuations is not as universal as stochastic electronic forces. This is because the effect of thermal vibrations is effectively reduced to a modification of the channel capacity which then becomes a random function of time with the statistical properties determined not only by the statistics of thermal vibrations but also the specific geometry of the channel. Through screening, the CPs are affected by shifts in relatively small regions of the configuration space, more precisely, in regions with linear dimensions smaller than the screening radius. Under these conditions, the finite correlation time of the shocks and the correlation function of random forces approximated by (2.3) can be neglected. In summary, we assume that the one-particle distribution of the CPs in transverse energy $P(\varepsilon_\perp, t)$ satisfies the Fokker–Planck equation describing the time course of the Markov process

$$\frac{\partial}{\partial t}P(\varepsilon_\perp,t) - L_{\mathrm{FP}}P(\varepsilon_\perp,t) = 0, \qquad (2.31)$$

where

$$L_{FP} = -a(\varepsilon_\perp)\frac{\partial}{\partial\varepsilon_\perp} + \frac{\partial^2}{\partial\varepsilon_\perp^2}b(\varepsilon_\perp).$$

Since channeling is a homogeneous process, $a(\varepsilon_\perp)$ and $b(\varepsilon_\perp)$ in (2.31) do not depend on time, and to find the solution of the Fokker–Planck equation it is necessary to use the method of separation of variables. Using this method, the solution can be written in the form of

$$P(\varepsilon_\perp,t) = \sum_{n=1}^{\infty} c_n \exp(-\lambda_n t)\phi_n(\varepsilon_\perp). \tag{2.32}$$

Here λ_n and $\varphi_n(\varepsilon_\perp)$ are the eigenvalues and eigenfunctions of the Sturm–Liouville differential operator L_{FP}

$$L_{FP}\phi_n(\varepsilon_\perp) = \lambda_n\phi_n(\varepsilon_\perp), \tag{2.33}$$

with the boundary condition

$$\phi_n(\varepsilon_\perp)\big|_S = 0.$$

The latter means the act of destruction of a dynamic system when the phase point reaches the absorbing boundary S. The eigenvalues λ_n are non-negative and satisfy $\lambda_1 < \lambda_2 < \lambda_3...$. The smallest eigenvalue λ_1 corresponds to a non-negative eigenfunction $\phi_1(\varepsilon_\perp)$.

Analysis of equation (2.32) gives a fairly complete picture of the evolution of the system, at least for asymptotically large times. Of course, after a long time in the sum (2.32) it is sufficient to retain a member, including the smallest eigenvalue λ_1, which characterizes the reciprocal lifetime $\lambda_1 = 1/\tau$. In this approximation we have

$$P(\varepsilon_\perp,t) = c_1 \exp(-\lambda_1 t)\phi_1(\varepsilon_\perp), \tag{2.34}$$

where the coefficient c_1 is determined from the condition of normalization of the distribution (in $\varepsilon_\perp$-space) on $1/\varepsilon_\perp^{cr}$.

2.4.2. Eigenvalues of the kinetic Fokker–Planck equation and the main functions of dechanneling theory

Charged particles moving in a crystal in the channeling mode are affected by forces due to the continuous potential of atomic chains (axial channel) or the atomic planes (plane channel), as well as the impact of electronic braking forces and random forces. The latter are associated with the random acts of interaction with the electrons (Section 2.2) and with the lattice atoms, carrying out the thermal vibrations. Under the influence of random forces the transverse energy of the CPs increases and may reach the critical value $\varepsilon_\perp^{cr}$. Reaching the value $\varepsilon_\perp^{cr}$ is seen as the annihilation of CPs at the absorbing boundary. Therefore, the boundary condition in the problem of dechanneling with (2.34) taken into account has the form

$$\phi_1\left(\varepsilon_\perp\right)\Big|_{\varepsilon_\perp = \varepsilon_\perp^{cr}} = 0. \tag{2.35}$$

It will be seen that the problem of dechanneling is a special case of a more general problem of the lifetime of the system exposed to random forces. Here we mean only 'soft' factors of influence, whose single effect only slightly changes the state of fast particles. In a general dechanneling problem one of the main partial tasks is to find the dechanneling length R_{ch}, which, of course, is proportional to τ_{ch} – the lifetime of the particles in the channeling mode. Of course, the approach that we discussed in the previous section can be fully utilized to calculate $R_{ch} = v_0 \tau_{ch}$. The generality of these results is determined by the fact that in the used approach no specific assumptions were made about the grade of the CPs or type of channel.

There are also effects which lead to dechanneling even for a single act of interaction. For example, scattering on impurity atoms, dislocations, etc. Description of such effects requires no introduction to the theory of the complex procedure of constructing the kinetic equation, in most cases it is sufficient to calculate the probability of single scattering. Single effects that lead to dechanneling are not considered in this book.

Thus, we assume that the distribution function of the CPs in the space of transverse energies satisfies (2.31), then the length of dechanneling and λ_1 – the smallest eigenvalue of L_{FP} – are related by

$$R_{ch} = v_0 \frac{1}{\lambda_1}.$$

To calculate λ_1, we will use the following model representation:

– A two-dimensional model is accepted in which the channel is formed by two atomic chains lying in the plane (x, z). Assuming that the potential of the channel is harmonic, it is also assumed that the CP is a one-dimensional harmonic oscillator, oscillating at frequency ω.

– Because the thermal motion of the atoms of chains is very slow, it is assumed that the CPs move in a 'frozen' two-dimensional lattice whose atoms are displaced from ideal positions of nodes. In addition, we assume that the rms amplitudes of displacements u are small, so that considerations can be restricted to members of the order of u^2.

– The interaction of fast particles with valence electrons is examined and only close collisions with large momentum transfers are considered. The resonant interaction of CPs with the collective mode of the electron gas provides a sufficiently small contribution to the kinetics of channeling and, therefore, the contribution of resonant interactions is not counted in the subsequent formulas.

Construction of the kinetic Fokker–Planck equation in the case of electron scattering was already discussed in Section 2.2.1, and there is no need to repeat it in the modified model. As for the Fokker–Planck equation in terms of particle scattering by thermal vibrations of the lattice atoms, one should take into account possible changes in channel width due to random displacements of atoms.

This effect is described by the random functions of time, which are included in the equation of transverse motion of particles between the walls of the channel. The result is a stochastic equation with the use of which in well-known methods of the theory of Markov processes [20] one can easily construct the desired equation of motion for the particle distribution function. In both cases, regardless of whether electron or thermal scattering is taken into account, we obtain an equation in the form coinciding with (2.31), but containing different kinetic coefficients

$$a(\varepsilon_\perp) = -\frac{1}{2}\varkappa, \qquad b(\varepsilon_\perp) = \frac{1}{2}\varkappa\varepsilon_\perp. \qquad (2.36)$$

Here

$$\varkappa = \begin{cases} 2\xi_x^{fl} v_0 & \text{(electron scattering)} \\ 2d\dfrac{1}{mv_0}\gamma|u|^2 & \text{(heat dissipation)} \end{cases}$$

$M\omega^2 = \sqrt{\gamma}$, and the force ξ_x^{fl} is defined by (2.12). Note also that $\omega^2 \sim U_a$, where U_a is the potential of the atom–atom interaction, which is calculated on the basis of the potential of atomic chains. Assessment of the square of frequency leads to the following conclusion

$$\varkappa \sim \gamma \sim |U_a|^2.$$

The main physical quantity in the theory of dechanneling is R_{ch} – dechanneling length. It is clear that the length R_{ch} is mainly determined by dynamic friction, which is included in the first term of the operator L_{FP} (2.31). As for the second term, it carries information about the slow spreading of the CPs on the levels of the transverse potential and plays the role of the correction in dechanneling.

On this basis, we consider the second term in L_{FP} as a perturbation and the eigenvalue λ_1 is written in the form of the perturbation

$$\lambda_1 = \lambda_1^{(0)} + \lambda_1^{(1)} + \lambda_1^{(2)} + ...,$$

where

$$\lambda_1^{(0)} = -2 \int_0^{\varepsilon_\perp^{cr}} d\varepsilon_\perp a(\varepsilon_\perp) |\phi_1(\varepsilon_\perp)|^2.$$

Complete calculation of λ_1 taking into account the diffusion process leads to rather complicated expressions [25], whereas in the zero approximation with allowance for (2.36) and boundary condition (2.35) we get

$$\lambda_{1(e)}^{(0)} = \frac{1}{mv_0 \, \varepsilon_\perp^{cr}} 4\pi Z_1^2 e^4 n_e L_e, \qquad \lambda_{1(th)}^{(0)} = \frac{1}{mv_0 \varepsilon_\perp^{cr}} \gamma |u|^2 d.$$

This, in turn, gives a simple expression for the dechanneling length

$$R_{ch}^{(0)} = v_0 \left(\lambda_{1(e)}^{(0)} + \lambda_{1(th)}^{(0)} \right) = 2E_0 \varepsilon_\perp^{cr} \left\{ 4\pi Z_1^2 e^4 n_e L_e + \gamma |u|^2 d \right\}^{-1}. \quad (2.37)$$

Thus, (2.37) expresses R_{ch} through parameters of the microscopic theory, and all the other typical functions of the dechanneling theory are expressed by R_{ch}. For example, the dechanneling function (normalized output) has the form

$$\chi(z) = 1 - q_1 \exp(-z / R_{ch}),$$

where q_1 is the survival probability of a particle at the entrance to the channel. R_{ch} can also be used to express the distribution function of the depth of 'knock-on' of the particles from the channel, the probability of finding the CPs at a predetermined depth and, finally, the profiles of the interstitial atoms injected into the crystal in the channeling mode. We note that even the most rough approximation of $\lambda \simeq \lambda_{1(e)}^{(0)}$ gives the result that reproduces the experimental data with good accuracy.

2.5. Three modes of motion of fast particles

2.5.1. The angular distribution of backscattered particles incident on thin crystals in the direction of chains or planes. Elastic collisions with atoms of the 'frozen chain'

Density in phase space σ (q, p, t) contains full information about the spatial distribution of particles and their distribution in momentum. In [31] the Liouville equation is derived for the CPs at the stage of their movement when inelastic scattering is still ineffective (small depth) and for understanding the evolution of the system it is sufficient to consider only the elastic processes. We use the result [31], but some clarifications will be made. Namely, we choose in a crystal a crystallographic direction with low indices and tightly packed rows of atoms along this direction. To study the motion of particles at small angles to these chains, the radius vector of the particle $\mathbf{q}$ and its velocity $\mathbf{v}$ is conveniently written as (ρ, z) and $(v_\perp, v_z)$, highlighting the longitudinal and transverse components. Excluding the collision equation of motion for the function σ $(\rho, v_\perp, z)$ in the accompanying coordinates has the form

$$
v_0 \frac{\partial}{\partial z}\sigma(\rho, v_\perp, z) + v_\perp \frac{\partial}{\partial \rho}\sigma(\rho, v_\perp, z) -
$$
$$
-\frac{1}{m}\frac{\partial U(\rho)}{\partial \rho}\frac{\partial}{\partial v_\perp}\sigma(\rho, v_\perp, z) = 0. \tag{2.38}
$$

In (2.38) the potential of atomic chains $U(\rho)$ is taken in the form of the standard Lindhard potential. This potential is uniform along the z axis and, therefore, is conserved in the absence of inelastic scattering of the transverse energy of the particles.

The CP subsystem can be regarded as a low-density gas, and to describe this gas it is enough to use the abbreviated description by the one-particle distribution. In order to obtain such a distribution, we must integrate $\sigma(\rho, v_\perp, z)$ for all variables except the variables of the selected particle. Note also that the passage of fast particles through a crystal at a certain depth of the order of the coherence length results in the establishment of a quasi-equilibrium stationary distribution of the CP.

Using the term quasi-equilibrium distribution, we would like to emphasize that we use the shortened description of the non-equilibrium state and apply the language of non-equilibrium statistical mechanics in the case when the whole system (CP plus thermostat) is far from thermodynamic equilibrium. It should be borne in mind that the quasi-equilibrium in the transverse configuration space occurs at much shallower depths of penetration than in the space of transverse momentum (velocity).

As shown in [32], in the quasi-equilibrium state the transverse coordinate and the transverse velocity of the particles are dedicated to the distribution function only in such a combination in which they appear in the expression for transverse energy $\varepsilon_\perp$. Therefore, the quasi-equilibrium stationary distribution of the CP depends only on $\varepsilon_\perp$. Omitting the details of the solutions of the Liouville equation (2.38) and the transition from the probability density $\sigma(\rho, v_\perp, z)$ to the one-particle distribution $g(v_\perp)$, we present the final result, which is applicable in the conditions of elastic collisions of CPs. So, we have

$$g(\varepsilon_\perp) = \frac{1}{\sigma^2} \int_{\rho_1}^{\rho_0} d\rho\, \rho \exp\left(-\frac{\rho^2}{2\sigma^2}\right) \times$$

$$\times \exp\left\{-\left(\frac{\tilde{d}}{\sigma}\right)^2 \left(\frac{\varepsilon_\perp - U(\rho)}{2E_0}\right)\right\} J_0\left(\xi(\varepsilon_\perp,\rho)\right). \tag{2.39}$$

In the formula (2.39) $\rho_0 = (\pi n_c d)^{-1/2}$ is the effective radius of the area attributable to a single atomic chain in the transverse plane, d is the distance between the nearest atoms in the chain, the lower limit of $\rho_1 = \rho_1(v_\perp)$ is determined from the condition $U(\rho_1) = \varepsilon_\perp$; $\tilde{d}$ is the distance from which a continuous potential of the chain acts on the particle [10]; σ is the standard deviation of the atoms of the 'frozen' chain of nodes; $J_0(x)$ is the Bessel function of zero order, the Bessel function argument is

$$\xi(\varepsilon_\perp,\rho) = i\rho\tilde{d}\sigma^{-2}\sqrt{\frac{1}{E_0}(\varepsilon_\perp - U(\rho))}.$$

The one-particle distribution (2.39) is normalized to unity.

Knowing the distribution (2.39), we can analyze the output of fast particles through a crystal surface in vacuum. Recall that the atomic chains, forming a channel, are 'frozen' and only a few atoms are shifted relative to the axis of the chain by σ. Let ψ be the angle between the velocity of the fast particle and the axis of the chain, and ρ' the distance from the axis of the particle in the transverse plane. In this model, the relation of statistical mechanics [33], establishing a connection between the distribution of transverse energies (2.39) and the angular distribution $I(\psi)$, is written in the form

$$I(\psi) = 2 \int_0^{\rho_0} d\rho' \rho' \frac{1}{\sqrt{\rho_0^2 - (\rho')^2}} g\left(E_0 \psi^2 + U(\rho')\right). \qquad (2.40)$$

The function $I(\psi)$ (2.40) characterizes the number of particles emerging from the crystal in vacuum with the direction of the velocity inside the solid angle $2\pi\psi \, d\psi$. We note that (2.40) determines the angular distribution at the level of description defined by non-equilibrium statistical mechanics. The possibility of further analytical calculations of the angular distribution is very limited. Therefore, we substitute (2.39) into (2.40) and calculate $I(\psi)$ by numerical integration. The results of numerical calculations are shown in Fig. 2.3. The calculation used the following parameters: $d = \tilde{d} = 1$, $\rho_0 = 7.2 \, a_0$, where a_0 is the screening radius included in the standard potential [10]. The figure shows the curves corresponding to three values of the dimensionless parameter $\eta = (d/2^{3/2}) \, (\psi_{cr}/\sigma)$:

1) $\eta = 1.4$;   2) $\eta = 1.0$;   3) $\eta = 0.37$.

Before presenting the physical picture that explains the features of function $I(\psi)$, it should be noted that the calculation of the angular distribution and its interpretation are given at different levels: in calculating the trajectories of the particles are not taken into account, while the interpretation considers assumptions on the trajectory to recreate the qualitative picture.

The complete energy diagram of the system in the channeling task can be taken from [34]. According to [34], penetration of the CP through the wall of the channel occurs by tunneling and for the interpretation of these transitions it is proposed to enter a particular zone with the energy adjacent to the top of the potential barrier – the quasi-channeling zone. This zone divides above-barrier and sub-barrier states or, in other words, separates the states that re energetically adjacent to the top of the barrier. Moreover, in the quasi-channeling regime the particles move in the channel region where the probability of close collisions with electrons and atoms of the channel wall, displaced from the nodes, is high enough.

In the harmonic approximation, the motion of CPs occurs in sinusoidal trajectories at the center of the channel (path a_1 in Fig. 2.4), the chaotic particles move on trajectories b_1, and the motion in the quasi-channeling regime is performed along the trajectories

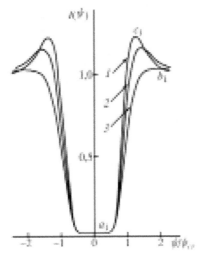

Fig. 2.3. Angular distribution of the particles exiting through the crystal surface into vacuum.

c_1. The motion of particles is accompanied by acts of penetration through the wall of the channel. For simplicity, we can assume that at the stage of motion when the particle 'slides' along the wall, its trajectory is localized within a shallow potential well, the lower level of which is adjacent to the top of the potential barrier (a graphic representation of this well will be given below).

Thus, the interpretation of the passage of fast particles used here is actually a 'hybrid' description of the system. On the one hand, the orbitals and the quasi-channeling zone are considered [34], i.e. physical quantities and concepts taken from the quantum theory, on the other – the trajectory – the subject of classical mechanics.

When interpreting the curve $I(\psi)$ in Fig. 2.3 it is necessary to take into account the fact that at small depths of penetration, i.e. depths of about 10^3 Å, three types of trajectories are possible, whose implementation depends on the angle of entry. Let us describe these paths:

– The particles, entering the channel at angles less than ψ_{cr}, undergo only sliding collisions with atoms of chains or planes, forming a channel. Moreover, in the immediate proximity of the chains (or planes) there is a region of width ~0.1 Å, into which the particle trajectories do not penetrate [14].

– If the angle of entry is much greater than ψ_{cr}, the particle is affected by lattice atoms, not as the impact of a continuous chain or plane. In this case, the fast particle moves at an arbitrary angle

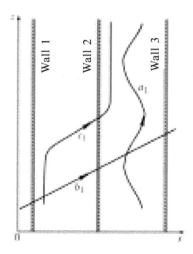

Fig. 2.4. Three types of trajectory of movement of particles in a flat channel.

to the chains, and its trajectory b_1 corresponds to the trajectory of the chaotic particles.

– If the angle of entry is close to ψ_{cr} (from the side of smaller values), the trajectory of a particle can take place in the immediate vicinity of the atomic chains, and the particle itself is subject to strong scattering. Schematically, such a trajectory is represented as trajectory c_1.

The yield of particles from the surface of the crystal in vacuum depends on the type of trajectory. Figure 2.3 shows three regions a_1, b_1 and c_1 of the curve $I(\psi)$, which correspond to three types of trajectories and, consequently, three modes of particle motion. In a_1 the output value is minimal, whereas in c_1 the output is much greater than that of the chaotic particles (area b_1). This is understandable, since the particles on the trajectories c_1 are subjected to the anomalously strong influence of the channel wall. Note also that the height of the parapet in region c_1 depends on the value of η and increases with this parameter. Mode motion of the particles that give output c_1 is analogous to quasi-channeling at great depths in the sense that these particles also fill the short-lived states energetically adjacent to the top of the potential barrier.

2.5.2. The energy distribution of backscattered particles. Elastic electron scattering

In calculating the angular distribution $I(\psi)$ (2.40), we took into account only elastic collisions with the atoms displaced from the nodes in the process of thermal motion. However, another formulation of the problem is possible, namely, to find a range of back-scattering particles in a thin crystal, taking into account elastic collisions with electrons. If the trajectory description is used, the motion of fast particles in this case can be divided into three stages.

In the first phase, the particles fly into the axial (plane) channel with initial energy E_0 and move to a depth of $z = R_1$ in the channeling mode. The energy losses at this stage are $S_{ch}(E) = B_0 S_r(E)$, where $S_r(E)$ is the loss of chaotic particles, E is the current value of the particle energy, B_0 is a numerical coefficient. In the second stage, the resonant transition of a particle takes place (resonance in the transverse energy) and, as a result, the particle goes into the quasi-channeling regime. Next, up to depth $z = R_2$ the particle moves along the trajectory of type c_1 (see Fig. 2.4) with the energy loss $S_{qch}(E) = B_1 S_r(E)$. Finally, the third stage begins with an act of backscattering of the particles, and then it goes the way of R_3 in the regime of chaotic motion and goes into the vacuum with energy E_3.

With the passage of the specified path $R_1 + R_2 + R_3$ under the influence of each elastic collision with an electron, there is some deviation of the particle from its initial trajectory. Such collisions are repeated many times and they differ in strength and direction of impact. In these circumstances, it is reasonable to abandon determinism in the description of particle motion, since we can not at any arbitrary point of time indicate the velocity of a particle and its position in space. The only correct way is to go into the world of probability.

Thus, we use the traditional method, by averaging the contribution of collisions in the process of moving a large number of macroscopically identical situations. Such averaging is essentially characteristic of the methods of statistical mechanics. Furthermore, deviating from the deterministic description of the system, we consider $E(t)$ and $E_3(t)$ as two normal Markov process and the energy distribution of particles in the three stages of movement P_1, P_2, P_3 can be written in Gaussian form. Then, the distribution of backscattered particles in elastic electron scattering takes the form

$$\tilde{I}(E_3,\psi) = N \int_{\min E_3}^{E_0} dE \frac{d\sigma(E,\psi)}{d\Omega} \int_{\min E_1}^{E_0} dE_1\, P_1(E_1) \times$$

$$\times \int_{\min E_2}^{E_1} d\bar{E}_2 P_2 (E - \bar{E}_2, R_2 - R_1) \times \qquad (2.41)$$

$$\times P_3 (E_3 - \bar{E}_3, R_3) \frac{1}{B_1 S_r (\bar{E}_2)}.$$

Here, $\bar{E}_2$ and $\bar{E}_3$ are the average energies of the particle in the second and third stages, the remaining symbols are standard.

Expression (2.41) contains a large number of physical parameters that can be determined experimentally. Therefore, if a particular experiment we can compare the distribution (2.41) with the observed spectrum of backscattering in a thin crystal and obtain a functional relationship of the two most interesting parameters B_0 and B_1. Comparison of [35] showed that the agreement of the energy spectra can be considered satisfactory if the values of B_0 are in the range $0.44 < B_0 < 0.48$. Comparison in [35] holds for protons with energies $E_0 = 2$ MeV, which initially moved in the axial channel of (111) silicon. The values of B_1 determined in this way and corresponding to two values of B_0, are presented in Table. 2.1. As can be seen from the data presented in the table, at shallow depths, where the inelastic (multiple) scattering particles is not effective, the condition $B_1 > 1$ is already satisfied. So the energy losses of quasi-channeled particles are greater than those of a random beam even in elastic electron scattering conditions. Of course, the loss of the CP is always less than the energy loss of chaotic particles, as evidenced by the condition $B_0 < 1$.

2.5.3. Singularities of the energy losses in the local theory

Let us continue the study of modes of motion of fast particles in crystals and differences of the energy losses when moving in different modes. To this end, we consider the energy loss, which gives the local theory that is applicable to crystals of sufficiently large thickness $L > 10^4$ Å.

The expression $a(\varepsilon_\perp)$ (2.30) is calculated by numerical integration and the resultant values of energy loss are represented graphically.

Table 2.1

B_0	0.44	0.46
B_1	1.1	1.07

As in Section 2.3.2, the calculation was carried out for helium ions with an energy of 1.5 MeV, moving in a planar channel of a (100) germanium crystal. The crystal temperature was 300 K.

As seen from Fig. 2.5, the function of the energy losses due to dynamic friction, $a(\varepsilon_\perp)$, is a maximum for $\varepsilon_\perp = U_0$. The presence of dynamic friction stronger than the friction of the particles in an amorphous medium explains the fact that particles with transverse energies, close to the top of the potential barrier U_0, move in the quasi-channeling mode. In this case, the particles move close to the atomic plane along the trajectories of the type c_I (see Fig. 2.4) and, of course, the scattering of particles due to interaction with atoms of the plane shifted from the ideal lattice sites is strongest.

It is interesting to compare the result of the local theory of energy losses with the corresponding result of the non-local theory [36]. In the non-local theory the loss function is expressed through the first moment of the transition probability in the space of transverse energies $W(\varepsilon_\perp, \varepsilon_\perp')$. Namely,

$$\mu_1(\varepsilon_\perp) = \int d\varepsilon_\perp' W(\varepsilon_\perp \varepsilon_\perp')(\varepsilon_\perp' - \varepsilon_\perp). \tag{2.42}$$

The integral transformation (2.30) of the diagonal matrix element of random effects (2.28) and the definition of the first moment of the probability (2.42) are profoundly different. Therefore, the behavior of $\mu_1(\varepsilon_\perp)$ is significantly different from the functional dependence

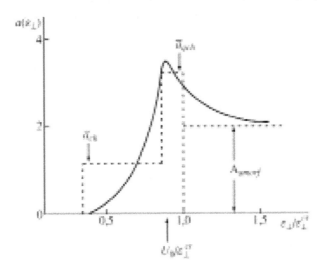

Fig. 2.5. Dependence of the energy losses on the transverse energy of the particles.

of $a(\varepsilon_\perp)$. Indeed, the moment $\mu_1(\varepsilon_\perp)$, due to nuclear scattering, is a smooth monotonically increasing function over the entire range of transverse energies $0 < \varepsilon_\perp < \varepsilon_\perp^{cr}$. At the same time, the function $a(\varepsilon_\perp)$ (see Fig. 2.5) has a feature in this range, namely, the maximum at $\varepsilon_\perp = U_0$. As for the larger values of transverse energy, with $\varepsilon_\perp > \varepsilon_\perp^{cr}$ it asymptotically approaches the value of losses in the amorphous medium A_{amorp}. It should be noted that the function $a(\varepsilon_\perp^{cr})$ was obtained here in the stochastic approach and is consistent with the function of the energy losses, calculated on the basis of the local Boltzmann equation (the results of the third chapter).

In each of the two transverse energy ranges, which correspond to channeling and quasi-channeling particles, the function $a(\varepsilon_\perp)$ is easy to average. After averaging, the loss function can be approximated by a piecewise linear function $\bar{a}(\varepsilon_\perp)$, which is represented by a dashed line in Fig. 2.5. With its two values, $\bar{a}_{ch}$ and $\bar{a}_{qch}$, the corresponding average loss of particles moving in the channeling and quasi-channeling mode, the following analytical expressions can be compared

$$S_{ch} = B_0 S_r, \qquad S_{qch} = B_1 S_r.$$

Analysis of these functions was given in the previous section. Because, according to Fig. 2.5, $\bar{a}_{qch} > \bar{a}_{ch}$, the output of the stochastic theory agrees qualitatively with the main result $B_1 > B_0$ [35].

The difference between the contributions of channeling and quasi-channeling can be demonstrated not only in experiments on the backscattering. The separation of the beam into fractions, as well as the difference of the energy losses, which comes from particles of different fractions, is clearly seen in the 'shoot-through' experiments. In particular, it was established in [37] that the particles with low energy losses, which fly into the crystal at small angles to the plane of the channel (near the center), provide a relatively larger number of particles passing through the crystal and transferred into vacuum through the back surface of the crystal plate.

The above considerations show that the particles of this fraction – the so-called well-channeled particles – move along the trajectories of type a_1 (see Fig. 2.4). At the same time, particles that enter the channel with large impact parameters, pass through the channel region with a high density of electrons and, in addition, are subjected to a strong effect of the fluctuations of the channel wall. They lose more energy and have a low normalized output. Because in the experimental conditions [37] the contribution of random particles to the angular distribution was subtracted, the fraction of the beam with high losses should be attributed to quasi-channeled particles moving along trajectories c_1.

3

Local Boltzmann kinetic equation as applied to planar and axial channeling

3.1. The distribution function for the complex of a fast particle and *r* atoms in the crystal

A separate independent direction has been formed in theoretical studies of non-equilibrium processes accompanying the passage of fast particles through a crystal – the local theory of channeling. In the works in this direction [11,12,38] special attention is paid to the derivation of local transport equations and the calculation on the basis of these equations of the kinetic characteristics of the system, i.e. diffusion functions and energy losses of particles.

However, as shown by analysis, these studies are in most cases of very limited interest, on the one hand, due to very simplified models, in which two constant and independent of the particle beams are fixed (channeled and chaotic), on the other hand – due to inadequate approximations used during derivations.

Apparently, a more consistent approach to solving problems of orientation effects of the local theory is proposed in this chapter. It is based on the ideas of Bogolyubov's kinetic theory [1]. This approach allows us to obtain a local kinetic equation for a rarefied gas of fast particles, interacting with lattice vibrations and valence electrons through the chain of BBGKY equations. And this, as will be shown, allows an analysis of kinetic effects, taking into account all possible modes of motion of particles.

Scattering of beam particles is the result of a large number of their collisions with atoms of the crystal. The crystal itself is a complex statistical system whose state is also changing during the interaction with the beam particles. Therefore, in the study of scattering of the particles by the crystal the particle beam and the crystal must be considered as a unified statistical system the various parts which vary in a coordinated manner and are interconnected with each other. In connection with this, the set of the beam particles and atoms in the crystal will be considered as a single two-component system.

Typically, the dynamics of the two-component system is described by the Boltzmann equation, obtained for a mixture of two gases. But for the beam–crystal system under consideration here the kinetic Boltzmann equation can not be used, firstly, because of the strong spatial inhomogeneity of the particle distribution function of the crystal particle and, secondly, due to the fact that we are interested mainly in those effects that are caused by both this strong spatial heterogeneity as well as strictly periodic arrangement of atoms in a crystal. Therefore, in this section, based on Bogolyubov's statistical equations [1] (more precisely, the BBGKY chain) we derive, for a two-component system, the kinetic equation for a rarefied gas of high-energy particles moving in a crystal, which would describe the effects associated with the periodicity and the strong spatial inhomogeneity of the distribution function of crystal atoms in the coordinates.

Consider a two-component system of N_g beam particles and N_c atoms of the crystal, which is contained in the volume Ω. Next, we introduce the distribution function dependent on time t

$$D\left(\mathbf{x}_1^g, \cdots, \mathbf{x}_{N_g}^g, \mathbf{x}_1^c, \cdots, \mathbf{x}_{N_c}^c; t\right)$$

of this system with respect to the dynamic states of the particle beam $\mathbf{x}_1^g, \ldots, \mathbf{x}_{N_g}^g$ in the crystal atoms $\mathbf{x}_1^c, \cdots, \mathbf{x}_{N_c}^c \left(\mathbf{x}_i^g = \left(\mathbf{q}_i^g, \mathbf{p}_i^g\right), \mathbf{x}_k^c = \left(\mathbf{q}_k^c, \mathbf{p}_k^c\right)\right)$. Distribution function $D\left(\mathbf{x}_1^g, \cdots, \mathbf{x}_{N_g}^g, \mathbf{x}_1^c, \cdots, \mathbf{x}_{N_c}^c; t\right)$ is a symmetric function with respect to permutations of the dynamical variables of each component separately. However, it satisfies the classical Liouville equation (1.2) (1.3).

Next, we introduce the distribution function of the complex of s fast atomic particles and r atoms in a crystal

$$\frac{1}{\Omega^s}\cdot\frac{1}{\Omega^r}F_{sr}\left(\mathbf{x}_1^g,\cdots,\mathbf{x}_s^g;\mathbf{x}_1^c,\cdots,\mathbf{x}_r^c;t\right)=$$

$$=\int_{\Omega_V}\cdots\int_{\Omega_V}d\mathbf{x}_{s+1}^g\cdots d\mathbf{x}_{N_g}^g\,d\mathbf{x}_{r+1}^c\cdots d\mathbf{x}_{N_c}^c\times$$

$$\times D\left(\mathbf{x}_1^g,\cdots,\mathbf{x}_{N_g}^g,\,\mathbf{x}_1^c,\cdots,\mathbf{x}_{N_c}^c;t\right),$$

where the integration is performed over the phase space Ω_V, i.e. space point $(\mathbf{q}_r^c,\,\mathbf{p}_r^c)$ and $(\mathbf{q}_s^g,\,\mathbf{p}_s^g)$.

We write the distribution function for the complex of a fast particle and r lattice atoms, using the 'coarsened' time scale on which the kinetic time is measured. According to [1], the distribution function of the complex

$$F_{1r}\left(\mathbf{x}_1^g;\mathbf{x}_1^c,\cdots,\mathbf{x}_r^c;t\right)$$

depends on time only through the temporal dependence of one-particle functions. In other words, according to the Bogolyubov's basic idea, F_{1r} is functionally dependent only on the distribution functions $F_{10}(t)$ and $F_{01}(t)$. We explain this fact in more detail.

The motion of fast channeled particles in the crystal is realized in terms of collective correlated interactions of charged particles with solids. In addition, when writing the distribution function of the complex of a single particle and r lattice atoms it is essential to take into account multiparticle correlations in a crystal lattice since a single atom can be involved in the movement of the collection of a large number of neighboring atoms. (A similar question was not formulated in the gas theory.) Given this, the correlation functions for the distribution of all possible groups of atoms in the lattice must be introduced in the distribution function of the complex in addition to $F_{10}(t)$ and $F_{01}(t)$. We get

$$F_{1r}=F_{1r}\left(\mathbf{x}_1^g,\mathbf{x}_1^c,\mathbf{x}_2^c\cdots\mathbf{x}_r^c\big|F_{10}(t),F_{01}(t),F_{02}(t),\cdots,F_{0r}(t)\right),$$

where

$$F_{10}(t)=F_{10}\left(\mathbf{x}_1^g,t\right),$$

$$F_{01}(t)=F_{01}\left(\mathbf{x}_1^c,t\right),$$

$$F_{02}(t)=F_{02}\left(\mathbf{x}_1^c,\mathbf{x}_2^c\big|F_{01}(t)\right),$$

$$\cdots\cdots\cdots\cdots\cdots\cdots\cdots\cdots$$

$$F_{0r}(t)=F_{0r}\left(\mathbf{x}_1^c,\mathbf{x}_2^c,\cdots,\mathbf{x}_r^c\big|F_{01}(t)\right).$$

Here, as usual [8], F_{10} and F_{01} functions satisfy the kinetic equation of the form

$$\frac{\partial F_{10}(t)}{\partial t} = A\left(\mathbf{x}_1^g \middle| F_{10}(t)\right).$$

As for the correlation functions of the distribution $F_{0s}(t)$ ($s = 2, 3,..., r$), which allow one to enter correlations into (3.1), then after on the basis of the Liouville equation a chain of equations is compiled for $F_{0s}(t)$; general solutions of these equations are not required, but only a partial solution of the form

$$F_{0s}(t) = F_{0s}\left(\mathbf{x}_1^c, \mathbf{x}_2^c, \cdots, \mathbf{x}_r^c \middle| F_{01}(t)\right).$$

Here it is appropriate to make two comments:
– Selection of a specified class of solutions is a general limitation of the Bogolyubov method. The distribution (3.1) for a binary system depends on the time in a functional manner, namely, through the time dependence of the two functions $F_{10}(t)$ and $F_{01}(t)$. This synchronization function (3.1) with the functions $F_{10}(t)$ and $F_{01}(t)$ is typical for a gas mixture with two kinds of particles [39] and binary systems in general.
– Selection of a specified class of solutions does not mean postulating the presence of real systems and specific conditions that correspond to this class of solutions. Not all systems and not under all conditions can they be described by the kinetic equation. In [1] determination in the kinetic regime corresponds to fulfilling all the conditions on the values of characteristic time intervals of different processes. Fulfillment of these conditions is associated ultimately with the nature of the interaction potential.
So, in the conditions of channeling of the fast particles the interaction potentials are such that the kinetic regime is realized. (Note also that the deviation of the crystal from equilibrium under the influence of the particles was rather weak. This condition is also feasible, and most theories are built without taking into account the channeling effect of particles on the crystal.) A detailed discussion of the existence of the kinetic regime was published in [40]. Note also that the problems of slow evolution of function F_1 and quick synchronization of F_2 will be discussed in detail in a simple example, namely, the example of a one-component system (see Section 3.7).

3.2. The chain of BBGKY equations in channeling theory

As has been noted, the theory of transport processes, including processes accompanying the motion of fast particles, requires a microscopic study. More precisely, justification of the level of microscopic description of a binary system: fast particles and the crystal. In some special cases, such justification can be provided using the fluctuation–dissipation theorem, which relates the spontaneous fluctuations and the macroscopic response of the system to external influences. However, the overall justification is given usually by means of kinetic equations, and it is appropriate to note that the method of the kinetic equation also requires deep motivation, which can be given on the basis of a chain of BBGKY equations.

Integrating the Liouville equation (1.2), (1.3) with respect to the dynamic variables N_g (s beam particles) and N_c (r atoms in the crystal) and the fact that the distribution function of D for the values of dynamic variables tends to infinity, converts to zero, we obtain the chain of BBGKY for the two-component system

$$
\frac{\partial F_{s,r}}{\partial t} = \left[H_{s,r}; F_{s,r} \right] + n_g \int dx_{s+1}^g \left[\sum_{1 \leq i \leq s} \Phi_{i,s+1}^g; F_{s,r+1} \right] +
$$

$$
+ n_g \int dx_{s+1}^g \left[\sum_{1 \leq k \leq r} \Phi_{s+1,k}; F_{s+1,r} \right] +
$$

$$
+ n_c \int dx_{r+1}^c \left[\sum_{1 \leq k \leq r} \Phi_{k,r+1}^c; F_{s,r+1} \right] +
$$

$$
+ n_c \int dx_{r+1}^c \left[\sum_{1 \leq i \leq s} \Phi_{i,r+1}; F_{s,r+1} \right], \tag{3.2}
$$

$$
\left(s = 0, 1, 2, \ldots; \quad r = 0, 1, 2, \ldots \right).
$$

Here, n_c is the density of atoms in a crystal; n_g is the density of fast particles

$$
H_{s,r} = \sum_{1 \leq i \leq s} \frac{\left(p_i^g \right)^2}{2m} + \sum_{1 \leq k \leq s} \frac{\left(p_k^c \right)^2}{2M} +
$$

$$
+ \sum_{1 \leq i < j \leq s} \Phi_{ij}^g + \sum_{1 \leq k < l \leq r} \Phi_{kl}^c + \sum_{\substack{1 \leq i \leq s \\ 1 \leq k \leq r}} \Phi_{i,k} \tag{3.3}
$$

– the Hamiltonian of a set of s particles in the beam and r atoms in the crystal (m – mass of the beam particles, M – the atomic mass of the crystal), and Φ_{ik}, Φ_{kl}^c, Φ_{ij}^g is the energy of interaction between the i-th particle of the beam and the k-th crystal atom, between the k-th and and l-th atoms in the crystal and between the k-th and l-th beam particles, respectively (for example, $\Phi_{ik} = \Phi\left(\left|\mathbf{q}_i^g - \mathbf{q}_k^c\right|\right)$).

When the density of beam particles n_g is so small that the interaction between the beam particles themselves is negligible compared with the interaction of beam particles with the atoms of the crystal, i.e. in the case of $n_g \rightarrow 0$, neglecting in the right-hand side of (3.2) terms proportional to n_g, the following system is obtained

$$\frac{\partial F_{s,r}}{\partial t} = \left[H_{s,r}; F_{s,r}\right] + n_c \int dx_{r+1}^c \left[\sum_{1 \le k \le r} \Phi_{k,r+1}^c; F_{s,r+1}\right] +$$

$$+ n_c \int dx_{r+1}^c \left[\sum_{1 \le i \le s} \Phi_{i,r+1}; F_{s,r+1}\right], \tag{3.4}$$

$$\left(s = 0, 1, 2,\ldots; \qquad r = 0, 1, 2,\ldots\right).$$

The chain of equations (3.4) shows that the distribution functions $F_{s,r}$ are defined through the distribution functiona $F_{s,r+1}$ ($s = 0, 1, 2,\ldots$; $r = 0, 1, 2,\ldots$). In particular, when $s = 0$ we see that $F_{0,r}$ is defined by $F_{0,r+1}$. Indeed,

$$\frac{\partial F_{0r}}{\partial t} = \left[H_{0r}; F_{0r}\right] + n_c \sum_{1 \le m \le r} \int dx_{r+1}^c \left[\Phi_{m,r+1}^c; F_{0,r+1}\right], \tag{3.5}$$

where, according to (3.3), we have

$$H_{0r} = \sum_{1 \le n \le r} \frac{\left(p_n^c\right)^2}{2M} + U_r^c, \qquad U_r^c = \sum_{1 \le m < n \le r} \Phi_{mn}^c,$$

$$H_{1r} = \frac{\left(p_1^g\right)^2}{2m} + \sum_{1 \le n \le r} \frac{\left(p_n^c\right)^2}{2M} + \sum_{1 \le n \le r} \Phi_{1n} + U_r^c,$$

$F_{0,r}$ is the r-particle distribution function of the crystal atoms and determines the state of the crystal itself. Equations (3.4) and (3.5) show that at a first approximation when the density of the beam particle beam is very small, in calculating the distribution function $F_{s,r}$ we can consider the distribution function of the crystal as given. This is due to the fact that at sufficiently low density of the particle beam the crystal state in the first approximation does not change. This is the consequence of the fact that at sufficiently low particle beam densities the state of the crystal to a first approximation does not change.

Our further goal is to obtain from the system (3.4) the kinetic equation for one-particle distribution function $F_{1,0}(\mathbf{x}_1^g)$ of the beam particle with respect to the coordinates and momenta, i.e. equations of the type

$$\frac{\partial F_{10}}{\partial t} = A\left(\mathbf{x}_1^g; F_{10}\right),\tag{3.6}$$

where $A(x_1^g; F_{10})$ is an expression that at any given time t is completely determined by the shape of the distribution F_{10} for the same time. Since the system of equations (3.4) implies that the distribution function $F_{s,r}$ is associated with $F_{s,r+1}$, then it is easy to see that for the equation of the form (3.6) it suffices to consider (3.4) only for values of $s = 0$ and $s = 1$, i.e. it suffices to consider equations of the form

$$\frac{\partial F_{0,r}}{\partial t} = \left[H_{0,r}; F_{0,r}\right] + n_c \int d\mathbf{x}_{r+1}^c \left[\sum_{1\leq k\leq r} \Phi_{k,r+1}^c; F_{0,r+1}\right],\tag{3.7}$$

$$\frac{\partial F_{1r}}{\partial t} = \left[H_{1r}; F_{1r}\right] + n_c \int d\mathbf{x}_{r+1}^c \left[\Phi_{1,r+1}^c; F_{1,r+1}\right] +\tag{3.8}$$

$$+ n_c \sum_{1\leq m\leq r} \int d\mathbf{x}_{r+1}^c \left[\Phi_{m,r+1}^c; F_{1,r+1}\right].$$

3.3. Local linearized Boltzmann equation

3.3.1. Planar channeling

To get an idea of the magnitude of the contribution made by a member of the right side of the system (3.4), (3.5) to change the distribution function $F_{s,r}$, we transform the system of equations (3.7), (3.8) to dimensionless form. The interaction potential Φ_{ik} is regarded as short-range (e.g. the potential of the type of the screened Coulomb interaction between an ion and a neutral atom, or Moliere potential) with short-range radius r_0 and the characteristic energy of interaction ε_0. Interaction between c-particles is also determined by the short-range potential (potential of the type of van der Waals forces) with short-range radius r_c and characteristic energy ε_c.

Then, introducing the characteristic values of the momenta for the given problem p_0^g and p_0^c (or velocities v_0^g and v_0^c)for the beam particles and atoms in the crystal, respectively, we define dimensionless variables by the relations

$$\tilde{q}_1^g = \frac{q_1^g}{r_0}, \quad \tilde{q}_r^c = \frac{q_r^c}{r_0}, \quad \tilde{p}_1^g = \frac{p_1^g}{v^g m},$$

$$\tilde{p}_r^c = \frac{p_r^c}{v^c M}, \quad \left(v^g = \dot{q}^g, \quad v^c = \dot{q}^c\right),$$

In the dimensionless equations (3.7) and (3.8) the integrals on the right side must be converted so that the main contribution to the integrals is provided by the dimensionless area of the order of unity. Then in the system of equations (3.7) and (3.8) we can distinguish a small parameter

$$\beta = n_c r_0^3 \left(\frac{\varepsilon_0}{2E_0}\right) \ll 1,$$

where ε_0 is the characteristic interaction energy.

However, a simpler way, avoiding unnecessary transformations of variables, can be used, namely, to consider β a formal parameter and enter it into the equation, in order to obtain information on the order of magnitude of each member. Then

$$\frac{\partial F_{1r}}{\partial t} = \left[H_{1r}'; F_{1r}\right] + \beta n_c \int dx_{r+1}^c \left[\Phi_{1,r+1}; F_{1,r+1}\right],$$

$$H_{1r}' = H_{1r} - U_r^c.$$
$$(3.9)$$

To obtain a closed equation for F_{1r} we use the expansion of function F_{1r} in powers of parameter β

$$F_{1r} = F_{1r}^{(0)} + \beta F_{1r}^{(1)} + \beta^2 F_{1r}^{(2)} + \cdots \cdots \qquad (3.10)$$

Substituting (3.10) into (3.9) for $r = 0$ we obtain the desired equation

$$\frac{\partial F_{10}}{\partial t} = \left[H_{10}; F_{10}\right] + \beta n_c \int dx_1^c \left[\Phi_{11}; F_{11}^{(0)}\right] +$$

$$+ \beta^2 n_c \int dx_1^c \left[\Phi_{11}; F_{11}^{(1)}\right] + \cdots \cdots \qquad (3.11)$$

To ensure that the initial condition corresponds to the actual physical conditions, we take into account the property of correlation weakening between some dynamic states. It is known that the particle is captured into the channeling mode only if it does not fit into the plane closer than the distance $r_0 + \sigma_\theta$. Here σ_θ is the average deviation of an atom of the crystal in a direction perpendicular to the crystallographic plane. Based on the results [1], it can be argued that, in a first approximation, a complex of one particle and r scattering centers moves as if it were isolated. Suppose that the complex considered at the initial state is in a state

$$S^{10}_{-\tau} x^g_1, \qquad S^{01}_{-\tau} x^c_1, \qquad S^{02}_{-\tau} x^c_2, \qquad \dots, \qquad S^{0r}_{-\tau} x^c_r,$$

where $(n_c r^3_0)^{-1}\tau_g \gg \tau \gg \tau_g, \tau_g = r_0/v^g$ is the passage time of the particle through the interaction region, $S^{10}_{\tau}, S^{01}_{\tau}, \dots, S^{0r}_{\tau}$ are the operators of the shift of the states by interval τ. In particular,

$$S^{11}_{-\tau} = \exp\left\{-\tau\left(\frac{1}{m}\mathbf{p}^g_1\frac{\partial}{\partial\mathbf{q}^g_1} + \frac{\mathbf{p}^c_1}{M}\frac{\partial}{\partial\mathbf{q}^c_1} - \frac{\partial\Phi_{11}}{\partial\mathbf{q}^g_1}\frac{\partial}{\partial\mathbf{p}^g_1} - \frac{\partial\Phi_{11}}{\partial\mathbf{q}^c_1}\frac{\partial}{\partial\mathbf{p}^c_1}\right)\right\},$$

$$S^{10}_{-\tau} = \exp\left\{-\tau\frac{1}{m}\mathbf{p}^g_1\frac{\partial}{\partial\mathbf{q}^g_1}\right\}, \qquad S^{01}_{-\tau} = \exp\left\{-\tau\frac{1}{M}\mathbf{p}^c_1\frac{\partial}{\partial\mathbf{q}^c_1}\right\}.$$

Since at the initial moment the particle was located at a distance greater than r_0 from the scattering center, the correlation between their dynamic states can be neglected. The correlation deviation is then negligible.

With this in mind and considering expansion (3.10), we write the initial condition for the system as

$$\lim_{\tau\to\infty} S^{1r}_{-\tau}\left\{F^{(0)}_{1r}\left(\mathbf{x}^g_1;\mathbf{x}^c_1,\dots,\mathbf{x}^c_r;S^{10}_{\tau}F_{10},\dots\right) - S^{10}_{\tau}F_{10}S^{0r}_{\tau}F_{0r}\right\} = 0,$$

$$\lim_{\tau\to\infty} S^{1r}_{-\tau}F^{(1)}_{1r}\left(\mathbf{x}^g_1;\mathbf{x}^c_1,\dots,\mathbf{x}^c_r;S^{10}_{\tau}F_{10},S^{01}_{\tau}F_{01},\dots,S^{0r}_{\tau}F_{0r}\right) = 0.$$
(3.12)

We write the derivative $\partial F_{10}/\partial t$ in (3.11) in variational derivatives $\delta F_{1r}/\delta F_{10}, \delta F_{1r}/\delta F_{01}, \delta F_{1r}/\delta F_{02},\dots$, and then substitute the expansion (3.10) into this equation and equate terms of equal powers of the parameter β. The result is a system of equations for the $F^{(0)}_{1r}, F^{(1)}_{1r}$, $F^{(2)}_{1r}$ functions defined as functionals of $F_{10}, F_{01}, F_{02},\dots$.

A method of solving equations for the $F^{(0)}_{1r}, F^{(1)}_{1r}$ functions with the initial conditions (3.12) was proposed in [1]. Using this method, we present a formal decision in the case of $i = 0$

$$F^{(0)}_{11}\left(\mathbf{x}^g_1,\mathbf{x}^c_1,t\right) = F_{10}\left(X^g_1\left(\mathbf{x}^g_1,\mathbf{x}^c_1\right),t\right)F_{01}\left(X^c_1\left(\mathbf{x}^g_1,\mathbf{x}^c_1\right),t\right). \quad (3.13)$$

The arguments of the function F_{01} are dynamic variables of the selected lattice atom interacting with a beam particle and all other lattice atoms

$$X_k^c\left(\mathbf{x}_1^g;\mathbf{x}_1^c,\cdots,\mathbf{x}_r^c\right)=\left\{P_k^c\left(\mathbf{x}_1^g;\mathbf{x}_1^c,\cdots,\mathbf{x}_r^c\right),Q_k^c\left(\mathbf{x}_1^g;\mathbf{x}_1^c,\cdots,\mathbf{x}_r^c\right)\right\},$$

$$P_k^c\left(\mathbf{x}_1^g;\mathbf{x}_1^c,\cdots,\mathbf{x}_r^c\right)=\lim_{\tau\to\infty}S_{-\tau}^{1r}\mathbf{p}_k^c, \tag{3.14}$$

$$Q_k^c\left(\mathbf{x}_1^g;\mathbf{x}_1^c,\cdots,\mathbf{x}_r^c\right)=\lim_{\tau\to\infty}\left\{S_{-\tau}^{1r}\mathbf{q}_k^c+\frac{1}{M}S_{-\tau}^{1r}\mathbf{p}_k^c\tau\right\}.$$

The variables X^g are determined by the same procedure. $F_{11}^{(1)}$ solution is more complicated than (3.13) [41]. Using these solutions, we can obtain a generalized local Boltzmann equation [41] linear in relation to F_{10}. The whole time dependence is included in the one-particle and two-particle distribution functions of the crystal F_{01} and F_{02}. Therefore, the equation [41] correctly describes the motion of fast particles only in the case when the characteristic evolution time of the subsystem of particles of the function the distribution functions of crystal atoms varies insignificantly. However, it does not take into account the specifics of the motion of particles in the channeling mode.

The feature of the motion of particles in the regime of planar channeling is that the motion of particles at small angles to the crystallographic plane is accompanied by a large number of correlated interactions in which the particle in each collision event only slightly deviates from the original direction.

To write the kinetic equation taking into account the sliding collisions, it is necessary, above all, to set an explicit form of the variables (3.14). To do this, using the equations of dynamics, we compute the coordinates and momenta of the particles after the collision, approximating the trajectory of the particles between two successive acts of direct collision. We get

$$X_1^g=\left\{\tilde{Q}_1^g,\tilde{P}_1^g\right\},\qquad X_1^c=\left\{\tilde{Q}_1^c,\tilde{P}_1^c\right\},$$

$$\tilde{Q}_1^g=q_1^g+\xi^g,\ \ \tilde{P}_1^g=p_1^g+\zeta^g,\ \ \tilde{Q}_1^c=q_1^c+\xi^c,\ \ \tilde{P}_1^c=p_1^c+\zeta^c,$$

$$\zeta^g=\int_0^\infty d\tau\nabla_{q^g}\Phi\left(\left|\mathbf{q}^g-\mathbf{q}^c-\left(\mathbf{v}^g-\mathbf{v}^c\right)\tau\right|\right),$$

$$\xi^g=\frac{1}{m}\int_0^\infty d\tau\,\tau\nabla_{q^g}\Phi\left(\left|\mathbf{q}^g-\mathbf{q}^c-\left(\mathbf{v}^g-\mathbf{v}^c\right)\tau\right|\right),$$

$$\zeta^g=-\zeta^c,\ \ \ \xi^g=-\left(\frac{M}{m}\right)\xi^c.$$

Bearing in mind the generalization of the equations to the case of electrons (as a subsystem of the thermostat), we assume that the parameter M can take two values: M_c – atomic mass of the lattice (thermal dissipation) and m_e – the mass of an electron (electron scattering), $\mathbf{v}^c$ and $\mathbf{v}^g$ – the velocity of the lattice atom and particle velocity.

If the collision term is calculated with an accuracy up to the second order with respect to the interaction potential, it suffices to replace the variables X^g and X^c, included in the collision integral, by x^g and x^c. Expanding then the classical Poisson bracket, which originally belonged to (3.11), we find the required kinetic equation – Boltzmann equation in Bogolyubov's form

$$\frac{\partial}{\partial t}F_{10}\left(\mathbf{x}^g,t\right)+\sum_i\frac{1}{m}p_i^g\frac{\partial}{\partial q_i^g}F_{10}\left(\mathbf{x}^g,t\right)=\left(\frac{\partial F_{10}}{\partial t}\right)_{st};$$

$$\left(\frac{\partial F_{10}}{\partial t}\right)_{st}=n_c\beta\sum_i\int d\mathbf{x}_1^c\frac{\partial}{\partial q_i^g}\Phi\left(\left|\mathbf{q}^g-\mathbf{q}_1^c\right|\right)\times$$

$$\times\frac{\partial}{\partial p_i^g}\left\{F_{10}\left(\tilde{Q}^g,\tilde{P}^g,t\right)F_{01}\left(\tilde{Q}_1^c,\tilde{P}_1^c\right)\right\}+$$

$$+n_c^2\beta^2\sum_{ij}\int d\mathbf{x}_1^c\frac{\partial}{\partial q_i^g}\Phi\left(\left|\mathbf{q}^g-\mathbf{q}_1^c\right|\right)\times$$

$$\times\int d\mathbf{x}_2^c\frac{\partial}{\partial p_i^g}\left\{\frac{\partial}{\partial p_j^g}F_{10}\left(\mathbf{x}^g,t\right)\left[F_{02}\left(\mathbf{x}_1^c,\mathbf{x}_2^c\right)-F_{01}\left(\mathbf{x}_1^c\right)F_{01}\left(\mathbf{x}_2^c\right)\right]\right\}\times$$

$$\times\int_0^\infty d\tau\frac{\partial}{\partial q_j^g}\Phi\left(\left|\mathbf{q}^g-\mathbf{q}_2^c-\left(\mathbf{v}^g-\mathbf{v}_2^c\right)\tau\right|\right),$$

(3.15)

(3.16)

where

$$i,j=1,2,3;\quad q^g=\left\{q_x^g,q_y^g,q_z^g\right\},\quad q_x^g\equiv q_1^g,\quad q_y^g\equiv q_2^g,\quad q_z^g\equiv q_3^g.$$

As for the stationary functions of the crystal F_{01} and F_{02}, then, given the proximity of the crystal to a state of thermodynamic equilibrium at temperature T, the functions can be represented as

$$n_cF_{01}\left(\mathbf{x}^c\right)=\tilde{F}_{01}\left(\mathbf{q}^c\right)\varphi\left(p^c\right),n_c^2F_{02}\left(\mathbf{x}_1^c,\mathbf{x}_2^c\right)=\tilde{F}_{02}\left(\mathbf{q}_1^c,\mathbf{q}_2^c\right)\varphi\left(p_1^c\right)\varphi\left(p_2^c\right).$$

Here

$$\varphi(p)=\left(2\pi MT\right)^{-3/2}\exp\left\{-\left(\frac{p^2}{2MT}\right)\right\}.$$

The normalization ratio for the one-particle distribution function has the form

$$\frac{1}{\overline{\omega}}\int d\mathbf{x}^g F_{10}\left(\mathbf{x}^g,t\right) = N_g, \qquad (3.17)$$

where $\overline{\omega}$ is the volume per one fast particle.

3.3.2. Axial channeling. The limiting case of a spatially homogeneous environment

So far we have considered the problem of planar channeling. We now discuss the question of how to change the equation in the transition to the axial channel. Consider one of the crystallographic directions in a crystal with small indices and packed chains of atoms that are located in this area. In studying the motion of particles at small angles to these chains it is convenient to represent the motion of particles as a superposition of longitudinal and lateral movements in relation to the direction of the chains. For this purpose we represent the radius vector of the particle in the form $\mathbf{q}$ $(\boldsymbol{\rho}, z)$, and its velocity $\mathbf{v}$ as $(\mathbf{v}_\perp, v_0)$, where $\boldsymbol{\rho}$ and $\mathbf{v}_\perp$ are the transverse radius vector and the velocity of a particle, respectively, $\mathbf{v}_\perp = (v_1, v_2)$, $\boldsymbol{\rho} = (\rho_1, \rho_2)$.

Repeating the procedure for deriving the kinetic equation based on the BBGKY chain, which is discussed in detail in Section 3.2, for the distribution function $F_{10}(\boldsymbol{\rho}, z, \mathbf{v}_\perp, v_0; t)$ in the case of the axial channel we obtain

$$\frac{\partial}{\partial t}F_{10} + v_0 \frac{\partial}{\partial z}F_{10} + \sum_i v_i \frac{\partial}{\partial \rho_i}F_{10} - \sum_i \frac{\partial}{\partial \rho_i}U_b(\boldsymbol{\rho})\frac{\partial}{\partial p_i}F_{10} =$$

$$= -\sum_i \frac{\partial}{\partial p_i}\left(A_i F_{10}\right) + \frac{1}{2}\sum_{ij}\frac{\partial}{\partial p_i}\left(B_{ij}\frac{\partial}{\partial p_j}F_{10}\right). \qquad (3.18)$$

Here, $(i, j) = (1, 2)$,

$$U_b(\boldsymbol{\rho}) = \sum_n \int d\boldsymbol{\rho}_c w\perp\left(\boldsymbol{\rho}_c - \boldsymbol{\rho}_{cn}\right)U_a\left(\left|\boldsymbol{\rho} - \boldsymbol{\rho}_c\right|\right)$$

– the average transverse potential of the set of atomic chains forming an axial channel, $U_a(\rho)$ – the continuous chain potential (1.7), $w_\perp(\boldsymbol{\rho}_c - \boldsymbol{\rho}_{cn})$ – the probability density distribution of atomic chains with a chain around the selected coordinate $\boldsymbol{\rho}_c$, $\boldsymbol{\rho}_{cn}$ – the radius vector of the n-th chain in the transverse plane. In addition, simplifying the form of equation (3.18), we have omitted in dynamic variable fast particles superscript g, replacing x^g by x.

As for the coefficients A_i and B_{ij}, in general they are quite cumbersome [42] and we write them only in the spatially homogeneous case. Indeed, if the distribution of atoms in the medium and the particle distribution is uniform, the equation for $F_{10}(\mathbf{v}_\perp, v_0; t)$ is greatly simplified. We have

$$\frac{\partial}{\partial t} F_{10}(\mathbf{v},t) = -\sum_i \frac{\partial}{\partial p_i}(a_i F_{10}) + \frac{1}{2}\sum_{ij} \frac{\partial}{\partial p_i}\left(b_{ij}\frac{\partial}{\partial p_j}F_{10}\right), \qquad (3.19)$$

where

$$a_i = \frac{n_c}{M_c}\int d\mathbf{q}' \frac{\partial^2 \Phi(|\mathbf{q}'|)}{\partial(q_i')^2}\int_0^\infty d\tau\, \tau \frac{\partial}{\partial q_i'}\Phi(|\mathbf{q}' - \mathbf{v}\tau|),$$

$$b_{ij} = 2n_c \int d\mathbf{q}' \frac{\partial \Phi(|\mathbf{q}'|)}{\partial q_i'}\int_0^\infty d\tau \frac{\partial}{\partial q_j'}\Phi(|\mathbf{q}' - \mathbf{v}\tau|).$$

Using the Fourier component of the potential of the atom–atom interaction $\Phi(k)$, it is easy to determine the vector a_i and tensor b_{ij} by simpler relations

$$a_i = \frac{1}{2M_c}\sum_j \frac{\partial}{\partial v_j}b_{ij},$$

$$b_{ij} = 2\pi^2 n_c \frac{v^2\delta_{ij} - v_i v_j}{v^3}\int_0^\infty dk\, k^3 |\Phi(k)|^2. \qquad (3.20)$$

Equation (3.18) can be applied to describe the motion of fast particles at small angles to the close-packed atomic chains. It makes it possible to take into account the spatial inhomogeneity of the distribution of atoms in a crystal, which is essential for the theory of orientation effects. As for equation (3.19), it would seem that it can be reduced to the kinetic equation for a weakly interacting gas. However, these equations cannot be compared directly because the coefficients in the equation for the weakly interacting gas (Landau, 1936) depend on the distribution function and, consequently, the equation is non-linear (in contrast to (3.19)).

Nevertheless, the difference between the equations can be greatly reduced if we restrict the consideration to examining the following physical situation. Let a homogeneous weakly interacting gas be in equilibrium at temperature T. Suppose that as a result of fluctuations at time $t = 0$ there is a small number of non-equilibrium particles, which are evenly distributed throughout the volume of the system.

Because of the low concentration, the non-equilibrium particles do not collide together and instead collide with the equilibrium particles, forming a thermostat, whose temperature is, of course, equal to T. The evolution of the subsystem of the selected particles can then be described by the equation for a weakly interacting gas, replacing the distribution functions that are included in the coefficient, by Maxwellian functions that are independent of time. As a result of this change the kinetic equation for the gas is linearized.

Now, without going into details, we can argue that in the weakly-interacting gas the evolution of non-equilibrium particles to equilibrium with the thermostat takes place by mechanisms known from the theory of Brownian motion. So, on the one hand, the non-equilibrium particles are affected by frictional forces, which in (3.19) correspond to the coefficient a_i, on the other hand – there is a significant velocity dispersion due to diffusion, which in (3.19) corresponds to tensor b_{ij}. It is worth noting that tensor b_{ij} (3.20) actually has the same form as the tensor in the equation for the gas.

Thus, the form of the linearized kinetic equation for a weakly interacting gas as applied to radiation physics is close to that of the equation (3.19), which allows us to estimate the contribution of dynamic friction and diffusion processes in the motion of particles in an amorphous solid.

3.4. The equation of local balance of the number of particles in the channeling problem

Hydrodynamic description is possible in the case where the processes of the transfer of energy, momentum and mass occur slowly over time and, moreover, are characterized by small spatial gradients of temperature, mass velocity and density. In this case, the state of the system can be described by a locally equilibrium distribution, i.e. Gibbs distribution, including parameters, depending on the coordinates and time. Such a description is usually given at the macroscopic level, however, the hydrodynamic approach can be used at the microscopic level description of the system [43].

As an example of the construction of nonequilibrium statistical mechanics on the hydrodynamic level of description, we describe derivation of the equation of the balance of the number of particles in the channeling theory. Namely, consider the field of particle number density $B(q_x, p_x, t)$, written in local form. Of course, field

$B(q_x, p_x, t)$ differs from the field of mass density, which usually appears in traditional fluid dynamics. The difference is the presence of a constant factor — the mass of the fast particle. The time dependence of the scalar field $B(q_x, p_x, t)$ is completely determined by the time course of the one-particle function

$$F_{10}(q_x, p_x; t) = g(q_x, p_x; t)\overline{\omega},$$

where $\overline{\omega}$ is the volume per one fast particle. The latter, as is known, is a local solution of the kinetic equation of the form (3.15). If the collision term of the kinetic equation is written in the operator form

$$\left(\frac{\partial F_{10}}{\partial t}\right)_{st} = L_1 g(q_x, p_x, t),$$

where L_1 is the collision operator, the density of the source of the field in a spatially inhomogeneous case can be written as

$$\sigma_B(q_x, p_x, t) = \int_{-\infty}^{\infty} dq_x' \int_0^{p_x} dp_x' \delta(q_x - q_x') L_1 g(q_x', p_x', t). \qquad (3.21)$$

With this definition of σ_B we examine each point of the configuration space in the x-direction with a delta function, and if there is no particle at point q_x, there is no contribution to the field source. It is worth recalling that in the case of a planar channel, we choose a coordinate system so that the x-axis is perpendicular to the crystallographic plane forming the channel wall.

Before disclosing the expression (3.21), it is necessary to simplify the form of the collision term (3.16). First, we assume that the particle distribution in y is homogeneous, because this corresponds to the real situation of planar channeling. Secondly, excluding from consideration the spreading of the particle beam along the y axis, we average the right side of (3.16) with respect to p_y. Thirdly, given that in the case of sliding impacts of a light particle moving in a lattice of heavy atoms, the values ζ^g, ζ^c, ξ^g, ξ^c are small quantities, we expand the collision term in powers of these variables and restrict ourselves to the senior members of the expansion. Using the the collision operator L_1 converted in this way, the source of the field B (3.21) is described by

$$\sigma_B(q_x, p_x, t) = \sigma_B^{(1)} + \sigma_B^{(2)} + \sigma_B^{(3)}, \qquad (3.22)$$

where

$$\sigma_B^{(1)} = \frac{1}{2} K(q_x) \cdot \left| \nabla_{p_x} g(q_x, p_x, t) \right|,$$

$$\sigma_B^{(2)} = \frac{1}{2} I(q_x) \cdot \left| \nabla_{q_x} g(q_x, p_x, t) \right|,$$

$$\sigma_B^{(3)} = R(q_x) g(q_x, p_x, t).$$

Here we use the notation

$$
\begin{aligned}
k(q_x) &\equiv K_{xx}(q) = 2 \int d\mathbf{q}^c \tilde{F}_{01}(\mathbf{q}^c) \nabla_{q_x} \Phi \left(\left| \mathbf{q} - \mathbf{q}^c \right| \right) \times \\
&\times \int_0^\infty d\tau \nabla_{q_x} \Phi \left(\left| \mathbf{q} - \mathbf{q}^c - (\mathbf{v}^g - \mathbf{v}^c)\tau \right| \right) + \\
&+ 2 \int d\mathbf{q}_1^c \, d\mathbf{q}_2^c \left\{ \tilde{F}_{02}(\mathbf{q}_1^c, \mathbf{q}_2^c) - \tilde{F}_{01}(\mathbf{q}_1^c) \tilde{F}_{01}(\mathbf{q}_2^c) \right\} \times \\
&\times \nabla_{q_x} \Phi \left(\left| \mathbf{q} - \mathbf{q}_1^c \right| \right) \int_0^\infty d\tau \nabla_{q_x} \Phi \left(\left| \mathbf{q} - \mathbf{q}_2^c - (\mathbf{v}^g - \mathbf{v}^c)\tau \right| \right).
\end{aligned}
$$

(3.23)

The functions $I(q_x)$ and $R(q_x)$, derived from (3.16), (3.21) and (3.22), are also quite complex. However, one should take into account that the functions included in $\sigma_B^{(n)}$ differ in their dependence on particle velocity v_0. Indeed, as shown by analysis,

$$K(q_x) \sim (v_0)^{-1}, \quad I(q_x) \sim (v_0)^{-2}, \quad R(q_x) \sim (v_0)^{-2}.$$

It follows that at high velocities in the expression (3.22) it is sufficient to retain the main term $\sigma_B^{(1)}$, including $K(q_x)$ (3.23), and the explicit form of $I(q_x)$ and $R(q_x)$ in this approximation is not required. At high temperatures, it is sufficient to write the distribution function of the lattice atoms $\tilde{F}_{01}(q^c)$ and $\tilde{F}_{02}(q_1^c, q_2^c)$ in a quasi-harmonic approximation [42]. Then, given the periodicity of unary and binary functions $\tilde{F}_{01}$ and $\tilde{F}_{02}$ with respect to q_x^c, they can be presented as a Fourier series and these expansions are substituted into (3.23). As a result, the diagonal element of matrix effects in the presence of 'thermal' scattering of fast particles takes the form

$$K(q_x) = K^{(1)}(q_x) - \Delta K^{(1)}(q_x),$$

$$K^{(1)}(q_x) = 2n_c 2l \sum_n B_n \int d\mathbf{q}^c \exp \left[i \frac{\pi n}{l} q_x^c \right] \times$$

(3.24)

$$\times \nabla_{q_x} \tilde{\Phi}\left(\left|\mathbf{q}-\mathbf{q}^c\right|\right) \int_0^\infty d\tau \nabla_{q_x} \tilde{\Phi}\left(\left|\mathbf{q}-\mathbf{q}^c-\mathbf{v}_0\tau\right|\right),$$

$$\Delta K^{(1)}(q_x) = 2n_c \, 2l \sum_n B_n^2 \int d\mathbf{q}_1^c \, d\mathbf{q}_2^c \, \exp\left[i\frac{\pi n}{l}\left(q_{1x}^c + q_{2x}^c\right)\right] \times \tag{3.25}$$

$$\times \tilde{n}_G\left(\rho_1^c - \rho_2^c\right) \nabla_{q_x} \tilde{\Phi}\left(\left|\mathbf{q}-\mathbf{q}_1^c\right|\right) \int_0^\infty d\tau \nabla_{q_x} \tilde{\Phi}\left(\left|\mathbf{q}-\mathbf{q}_2^c-\mathbf{v}_0\tau\right|\right). \tag{3.26}$$

Here

$$B_n = \frac{1}{2l} \exp\left[-\left(\frac{\pi n}{l}\right)^2 \frac{\sigma_T^2}{2}\right],$$

$\tilde{n}_G$ is the normal distribution with variance $2\sigma_T^2$. In the expressions (3.25) and (3.26) the interaction potential of the fast particle with the lattice atom $\tilde{\Phi}$ is represented by the Moliere approximation of the Thomas–Fermi potential V_M.

Since the velocity of thermal motion of lattice atoms is much smaller than particle velocity v_0, we can assume that the particle is moving fast in the frozen lattice of atoms with a fixed scatter of the atoms σ_T. Under these conditions, to reconstruct the full physical picture, it is necessary to introduce a static form factor $S(q_1, q_2)$, defining the spatial correlation between the lattice atoms. Then, using the factor $S(q_1, q_2)$, the diagonal element of the matrix of effects in the case of scattering of fast particles on the thermal displacements of the lattice atoms can be written in full form and in symmetrized form [41].

Expressions (3.22), (3.24) – (3.26) give the final expression for the density of the field source $\sigma_B(q_x, p_x, t)$. If we neglect the processes in the momentum space and additionally introduce the flux of particles $J_B(q_x, p_x, t)$ to the theory, then the local balance equation describing the rate of change of the field B takes the form

$$\frac{\partial}{\partial t} B(q_x, p_x, t) = -\nabla_{q_x} J_B(q_x, p_x, t) + \sigma_B^{(1)}(q_x, p_x, t). \tag{3.27}$$

Let us explain the physical meaning of the individual terms of this equation in the case of channeling of charged particles. For this we integrate both sides of (3.27) at $\sigma_B^{(1)} = 0$ with respect to the volume between the two crystallographic planes, forming a channel. Then, using Gauss' theorem, we find that changes in the density of particles in the selected volume are equal to our chosen particle flux density across the surface of this volume. If the source field

$\sigma_B^{(1)}(q_x, p_x, t)$ (3.22) is not zero, it provides an additional contribution to the change in the number density of particles. We emphasize that these changes are not associated with the flux of particles through the boundary of the volume, in other words, they are not associated with the release of particles from the channeling regime. The yield of particles through the surface of the selected volume can lead to a transition to the chaotic part of the beam.

Here it is appropriate to mention one of the applications of the general equation of field theory (3.27). Specifically, the wording of the second law of thermodynamics, using (3.27). According to the second law, the entropy of a closed system in any spontaneous process should increase and this increase is suspended only when the system reaches a static equilibrium. (In equilibrium, the entropy is maximum). Of course, the rate of change of entropy $\tilde{S}(q_x, p_x, t)$, calculated for a single particle, is described by the equation of the same species as the general equation of local balance (3.27). Given the entropy flux $\tilde{J}_S$ we have

$$\frac{\partial}{\partial t}\tilde{S}(q_x, p_x, t) = -\nabla_{q_x}\tilde{J}_S(q_x, p_x, t) + \sigma_S(q_x, p_x, t). \qquad (3.28)$$

Since the entropy flux $\tilde{J}_S$ in a closed system is always positive, the source of entropy, according to (3.28), satisfies $\sigma_S \geq 0$. Confirmation of the sign of the entropy of the source is also the formulation of the second law of thermodynamics. In turn, the thermodynamic law of increasing entropy is a mathematical expression of the irreversibility of the macroscopic non-equilibrium processes.

3.5. Diagonal element of the matrix of effects in electron scattering

Let us study the motion of particles in electron scattering. In the process of scattering on electrons a significant role is played by temporal correlations. Now we need to take into account the spatial and temporal correlations when the value of the electron density ρ at the point $\mathbf{q}$ at time t affects the density at the point $\mathbf{q}'$ at time $t + \tau$. The dynamic form factor $S(\mathbf{q}, \mathbf{q}'; \tau)$, characterizing the spatial-temporal correlation between the electrons, can be expressed in terms of the two-time retarded density–density Green function

$$\left\langle\!\left\langle \rho\!\left(\mathbf{q}',t+\tau\right)\rho\!\left(\mathbf{q},t\right)\right\rangle\!\right\rangle = \theta(\tau)\frac{1}{i}\left\langle\left[\rho\!\left(\mathbf{q}',t+\tau\right),\rho\!\left(\mathbf{q},t\right)\right]\right\rangle_0, \qquad (3.29)$$

if the system of electrons weakly deviates from the statistical equilibrium. (Green's function method is discussed in Section 4.3.) When writing Green's function, we used the standard notation: $\langle...\rangle_0$ – averaging over the equilibrium distribution of electrons. The Fourier component of the dynamic form factor with respect to the spatial and temporal variables should be related to the imaginary part of the Fourier component of Green's function

$$S\left(\mathbf{k},\mathbf{k}';\omega\right)=\frac{1}{\pi}\mathrm{Im}\left\langle\!\left\langle \rho_{\mathbf{k}}\big|\rho_{\mathbf{k}'}\right\rangle\!\right\rangle_\omega,$$

where $\rho_{\mathbf{k}}$ is the Fourier component of density fluctuations of the electron gas. It should be emphasized that after defining $S(\mathbf{k}, \mathbf{k}'; \omega)$ we switched to a 'hybrid' description, where the particles are still treated as a classical subsystem, and electrons – as a quantum one.

The diagonal element of the matrix of the effect of electrons on fast particles $\tilde{K}_{xx}(q)\equiv\tilde{K}(q_x)$ is easy to obtain, using the diagonal element $K_{xx}(q)\equiv K(q_x)$ (3.24) in the case of scattering by the displaced lattice atoms. It is enough to replace $\tilde{\Phi}(q)$ in (3.25) and (3.26) by $\Phi_e(q)$ – the Fourier component of the interaction potential of the fast particle with an electron gas. Before we can use the general form of the dynamic form factor $S(\mathbf{k}, \mathbf{k}'; \omega)$, we decompose the diagonal matrix element $\tilde{K}(q_x)$ into a Fourier series. The Fourier component of $\tilde{K}(q_x)$ is denoted $A_2(h_x)$. Then

$$A_2\left(h_x\right)=\int d\omega\frac{1}{v_0}\int d\mathbf{k}\,R_2\!\left(-\mathbf{k}+\frac{h_x}{2},\mathbf{k}+\frac{h_x}{2};\,\omega\right)\times$$

$$\times\delta\!\left(\frac{\omega}{v_0}+k\cos\tilde{\theta}\right), \qquad (3.30)$$

$$R_2\left(\mathbf{k},\mathbf{k}';\omega\right)=-S\left(\mathbf{k},\mathbf{k}';\omega\right)k_x k_x'\,\Phi_e\left(\big|\mathbf{k}\big|\right)\Phi_e\left(\big|\mathbf{k}'\big|\right).$$

It follows from (3.30), taking into account the space–time correlations (in (3.25) and (3.26), we considered only the spatial correlations) provides additional integration over the transferred energies. Of course, in the integration over the transferred energy and momenta we also consider the conservation law $\delta(\omega + \mathbf{k}\mathbf{v}_0)$. In (3.30), this law ($\tilde{\theta}$-angle between $\mathbf{k}$ and $\mathbf{v}_0$) is written in such a way that the dependence of A_2 on the rate (in cases of high velocities) is allocated explicitly. Indeed, the analysis shows that at $kv_0 \gg \omega$ in the integral over dk (3.30) it is sufficient to consider only the 'older' term, and the latter does not depend on v_0, so that $A_2(h_x) \sim (1/v_0)$.

$\tilde{K}(q_x)$ will be transformed. First, restricting ourselves to second order terms with respect to the screened Coulomb potential Φ_e, Green's function in (3.29) is replaced by Green's function of zero order in the interaction. Second, in line with traditional notions of the theory of channeling, we assume that the electron gas is isotropic. In this case, the imaginary part of the density–density Green function depends only on the modulus $|\mathbf{k}|$. Taking this into account, integration in (3.30) is performed over angular variables. Third, in integration in (3.30) over ω, we use the sum rule for the retarded Green function [44]. The result is

$$\tilde{K}(q_x) = \frac{1}{v_0}\left(Z_i e^2\right)^2 \sum_{h_n \geq 0} B^{(2)}(h_n)\left\langle \rho_{h_n} \right\rangle^{(0)} \cos(h_n q_x),$$

$$B^{(2)}(h_n) = \frac{1}{n_0}\int_{k_c}^{\Lambda_2} dk\left[k^3 - \frac{1}{4}kh_n^2\right] \times \tag{3.31}$$

$$\times \left[\gamma_n^2(k) - k^2 h_n^2\right]^{-1}\{n_0 - \mathcal{P}_n(k)\},$$

where $\gamma_n(k) = k^2 + \varkappa^2 + (1/4)h_n^2$; $\Lambda_2 = 2m_e v_0$, $\langle \cdots \rangle^{(0)}$ – averaging over the equilibrium state of non-interacting electrons,

$$\mathcal{P}_n(k) = \left\langle \rho_{h_n + k}\rho_{-k}\right\rangle^{(0)} \frac{1}{n_{loc}(h_n)},$$

$$n_0 = \left\langle \rho(h_n = 0)\right\rangle^{(0)}, \quad n_{loc}(h_n) = \left\langle \rho_{h_n}\right\rangle^{(0)}.$$

The minimum value of k in the integral (3.31) is defined as $k_c = \omega_p/v_F$, since for the processes of diffusion type it is sufficient to consider the contribution of close collisions with the individual excitations of the electron gas.

The same model representations are used in all studies known to us of diffusion dechanneling (see, for example, [45]). On the basis of (3.30) and (3.31), the unknown quantity can be written as follows: $\tilde{K}(q_x) = K^{(2)}(q_x) - \Delta K^{(2)}(q_x)$, where the first term, including n_0 in the brace in $B^{(2)}$ gives the term $K^{(2)}(q_x)$, and the second term, which includes $\mathcal{P}_n(k)$, gives a correlation part of the function – $\Delta K^{(2)}(q_x)$.

We find the final form of $\tilde{K}$ in electron scattering. First of all, we restrict calculation to calculating $B^{(2)}(h_n)$ in the random-phase approximation because analysis of the equation in this approximation is greatly simplified in comparison with analysis in the Hartrey–Fock approximation. So, we substitute in $B^{(2)}$ the expression for $\mathcal{P}_n(k)$ in the random-phase approximation and perform integration over the variable k. Next, using the Poisson equation, we calculate the density of the local valence electrons $n_{loc}(q)$ in accordance with how it was done in [46]. If in the expression for the plamar potential, included in the Poisson equation, we restrict outselves to the fourth-order anharmonicity limit

$$U_{pl}(q_x) = k_1 (q_x)^2 + k_2 (q_x)^4,$$

we get

$$n_{loc}(q_x) = n_c Z_{loc}^{(2)}(q_x)^2 = \frac{3k_2}{\pi e}(q_x)^2,$$

where $Z_{loc}^{(2)}$ is the number of local electrons in the atom lattice. Accordingly, the density of the homogeneous electron gas is

$$n_0 = n_c Z_{loc}^{(2)} = \frac{2k_1}{2\pi e}.$$

Given the explicit form of $B^{(2)}$ (h_n), n_0 and $n_{loc}(q_x)$ (more precisely, the Fourier components of the function $n_{loc}(q_x)$), summation with respect to h_n in $\tilde{K}(q_x)$ (3.31) is carried out. It is known that in the case of high particle velocities the screening effect contribution to the kinetic coefficients is negligible. Therefore, neglecting for simplicity screening, the final expression for $\tilde{K}(q_x)$ can be written as

$$K^{(2)}(q_x) = \frac{1}{v_0}(Z_1 e^2)^2 \left\{ n_c Z_{loc}^{(0)} \ln\left(\frac{2m_e v_0 vF}{\omega_p}\right) \right\} +$$

$$+ \frac{1}{2} n_c Z_{loc}^{(0)} \ln\left| \frac{(2m_e v_0)^2}{9(\pi/4l)^2 - k_c^2} \right| (2l/\pi)^2 u^2 \right\}, \qquad (3.32)$$

$$\Delta K^{(2)}(q_x) = \frac{1}{v_0}(Z_1 e^2)^2 n_c Z_{loc}^{(2)} \frac{8}{3}\left(\frac{4l}{\pi}\right)^2 f(u),$$

where

$$u = \frac{\pi}{2}\frac{q_x}{l}, \qquad f(u) = \frac{1}{3}\cos u - \frac{\pi}{8}\cos^2 u.$$

The value of $K_{xx} = K(q_x)$ (3.24)–(3.26) represents a diagonal element of the force matrix. The results of numerical calculation

(3.24) in terms of phonon scattering are shown in Fig. 3.1a. Here curve μ_1 is the function $K^{(1)}(q_x) / K_3$,

$$\mu_2 = \Delta K^{(1)}(q_x) / K_3, \, \mu_3 = \frac{K_{xx}}{K_3}, \, K_3 = \frac{8\pi}{v_0} n_c \left(Z_1 Z_2 e^2\right)^2.$$

The calculations were performed for helium ions with an energy of 1.5 MeV in the planar channel (100) of a germanium crystal at $T = 300$ K. As seen from Fig. 3.1a, K_{xx} takes the form of a Gaussian curve, which is due to thermal smearing of the atoms of the plane. The correlation part – the μ_2 curve – gives a negative correction (the same picture in the case of electron scattering) in the diagonal element of the full matrix

$$K_{xx} = K^{(1)}(q_x) - \Delta K^{(1)}(q_x).$$

Graphs of the matrix of the effects in electron scattering (3.32)

$$\tilde{K}_{xx} = K^{(2)}(q_x) - \Delta K^{(2)}(q_x)$$

at zero temperature are shown in Fig. 3.1b, where the curve μ_4 is the function $K^{(2)}(q_x)/K^{(2)}(0)$, $\mu_5 = \Delta K^{(2)}(q_x)/K^{(2)}(0)$, $\mu_6 = \tilde{K}/K^{(2)}_{(0)}$. Calculations μ_4, μ_5, μ_6 performed for protons with energies of 0.5 MeV, moving in a planar channel of the silicon crystal.

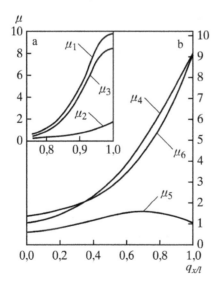

Fig. 3.1 A diagonal element of the force matrix in terms of phonon (a) and electron (b) scattering.

In contrast to the scattering by thermal lattice vibrations, the effect of electrons is considerable across the width of the channel, including the center channel. It should be noted that the curves in Fig. 3.1b refer to the case of channeling of light atomic particles of energy $E_0 \sim 1$ MeV, since in deriving (3.32) we took into account only the contribution of valence electrons. Naturally, at higher energies the contribution of the electrons of the atomic cores must also be taken into account.

3.6. Features of the kinetic functions as a consequence of locality of the theory

The balance equation for the particle number density (3.27) includes, as evident from (3.22), the diffusion flux in the p_x-representation

$$\frac{1}{2}\left\{K^{(1)}(q_x) - \Delta K^{(1)}(q_x)\right\}\nabla_{p_x} g(q_x, p_x, t).$$

Based on the definitions (3.24)–(3.26), using the specified flux, we can provide a complete description of the diffusion process either in the space of transverse momenta or in the $\varepsilon_\perp$-space.

Taking into account the well-known relationships presented in Section 2.3.3, the diffusion function (in dimensional form) can be written in the $\varepsilon_\perp$-space. We have

$$b(\varepsilon_\perp) = \frac{1}{m} \cdot \frac{1}{\varkappa(\varepsilon_\perp)} \int dq_x \left(K^{(1)}(q_x) - \Delta K^{(1)}(q_x)\right) \times$$
$$\times \left[\frac{1}{m}\left(\varepsilon_\perp - U_{pl}(q_x)\right)\right]^{1/2}. \tag{3.33}$$

However, complete study of the motion of fast particles in crystals requires not only study of the diffusion process in the space of transverse energies, but also consideration of the energy loss in inelastic scattering. To do this, we could derive a local balance equation for momentum density and then, based on it, to find the explicit form of the friction coefficient.

However, it is much easier to use a common expression for the coefficient $a(\varepsilon_\perp)$ provided by the stochastic theory. Indeed, comparing (3.32) with $b(\varepsilon_\perp)$ (2.30) shows that in the transition to the formalism based on the Boltzmann equation, the kernel of the integral transformation should take the value of $K^{(1)}(q_x) - \Delta K^{(1)}(q_x)$. Consequently, the function of the energy losses (in dimensional form), according to (2.30), can be written as follows

$$a(\varepsilon_\perp) = \frac{1}{2m} \frac{1}{\varkappa(\varepsilon_\perp)} \int dq_x \left(K^{(1)}(q_x) - \Delta K^{(1)}(q_x) \right)$$

$$\times \left[\frac{1}{m}(\varepsilon_\perp - U_{pl}(q_x)) \right]^{-1/2}. \qquad (3.34)$$

Using methods of numerical integration, we calculate the diffusion function (3.33) in the case of thermal scattering at three temperatures $T_1 = 300$ K, $T_2 = 500$ K, $T_3 = 700$ K. In the calculation of $b(\varepsilon_\perp)$ we used (3.25) and (3.26) (including the curves in Fig. 3.1a), as well as the expression (3.33).

The results of numerical calculation of the diffusion function in the case of α-particles with the energy of 1.5 MeV, moving in the channel of (100) germanium, are shown in Fig. 3.2, where the following notation is used:

$$y_i(\varepsilon_\perp) = \frac{b(\varepsilon_\perp)}{D_1}, \quad D_1 = 4\pi n_c \left(Z_1 Z_2 e^2 \right)^2 \frac{2\varepsilon_\perp^{cr}}{m v_0}, \quad X_i = \frac{U_0(T_i)}{\varepsilon_\perp^{cr}},$$

(where $i = 1, 2, 3$); $U_0(T)$ is the top of the potential barrier of the plane at temperature T; $\varepsilon_\perp^{cr} = E_0 \psi_{cr}^2$ is the critical transverse energy.

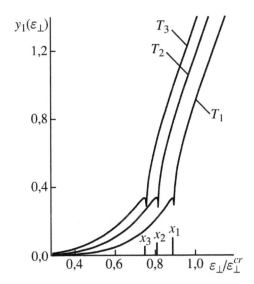

Fig. 3.2 The diffusion function of particles at three lattice temperatures.

Recall that, according to the original version of Lindhard's classical theory, at angles less than critical ψ_{cr}, the distribution of fast particles is divided into two parts: channeled and chaotic. Channeled particles move in the center of the channel, making periodic reflections from crystallographic planes, which form its walls. As for the chaotic part of the beam particles, they move along straight paths at arbitrary angles to the plane of the channel. In Fig. 3.2 these parts of the beam correspond to two areas: $\varepsilon_\perp < U_0(T)$ – channeled particles, $\varepsilon_\perp \geq \varepsilon_\perp^{cr}$ – chaotic ones. The lowest values of the diffusion function correspond to the channeling regime. Furthermore, Fig. 3.2 shows that in the case of random motion the function $b(\varepsilon_\perp)$ increases linearly with increasing $\varepsilon_\perp$.

But the most interesting result is observed (see Fig. 3.2) for $\varepsilon_\perp \sim U_0(T)$. Indeed, the diffusion function has a kink at $\varepsilon_\perp = U_0(T)$. The point is that the state of the particles, which are energetically adjacent to the top of the barrier, corresponds to the quasi-channeling zone [34]. In other words, another mode of motion is possible in this area of the transverse energies – quasichanneling – which is not considered in the classical Lindhard theory of channeling [10]. In the quasi-channeling regime, particles move along the crystallographic plane, making random transitions from one wall to the next (if the crystallographic planes are numbered from left to right, then – the transition from the first wall to the second, after a certain period of time – from second to third, etc.).

Local theory can take into account the difference in contributions provided by the particles moving at great depths in the channeling, quasi-channeling and chaotic motion regimes: hence the complex functional relationship, including the appearance of features (inflection) of the function $b(\varepsilon_\perp)$ forms. At the same time, non-local theory (see, for example, [47]) gives smooth, monotonically increasing functions $\varepsilon_\perp$ which have no singularities. Note also that in the region $U_0(T) < \varepsilon_\perp < \varepsilon_\perp^{cr}$ (see Fig. 3.2) transition takes place from quasi-channeling to chaotic motion; as shown by analysis, the trajectories of the quasi-channeled particles with increasing $\varepsilon_\perp$ change continuously into direct trajectories corresponding to particles of a chaotic beam (trajectories b_1 in Fig. 2.4).

Figure 3.2 shows a parametric dependence of the diffusion function of the temperature of the crystal. It is clearly seen that the channel potential barrier $U_0(T)$, obtained by numerical integration, decreases with increasing temperature. This effect is in satisfactory agreement with the results of analytical calculation of the vertices of the potential barrier [48], where the plane of the channel is viewed as a continuous set of harmonic oscillators, and the potential of the plane was calculated by adding up the individual layers formed by the oscillators.

The dimensionless energy losses

$$y_2(\varepsilon_\perp) = a(\varepsilon_\perp)\{\varepsilon_\perp^{cr}/D\}$$

in the same physical situation as in the calculation of $b(\varepsilon_\perp)$ are shown in Fig. 3.3 for three temperatures – 300, 500 and 700 K.

According to Fig. 3.3, the lowest energy losses $a(\varepsilon_\perp)$ (3.34) correspond to the CPs, i.e. the values of the transverse energy $\varepsilon_\perp < U_0$, since the motion in the channeling mode is at the center of the channel (path a_1 in Fig. 2.4). The function $a(\varepsilon_\perp)$ reaches its maximum at $\varepsilon_\perp = U_0$, reflecting a significant increase in energy loss for particles that fill the state, adjacent to the top of the energy potential barrier of the channel U_0. This is understandable, since in quasi-channeling the particles move near the atomic planes. The movement of these particles has a complex quasi-periodic nature, as is evident from Fig. 2.4 (path c_1). As for the energy losses of a chaotic beam, according to Fig. 2.4, with $\varepsilon_\perp > \varepsilon_\perp^{cr}$ they asymptotically approach a constant value y_{amorp}, corresponding to the dimensionless losses in the amorphous medium.

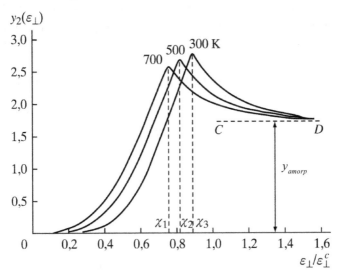

Fig. 3.3. Energy losses particles at three lattice temperatures. CD line – asymptotics at large $\varepsilon_\perp$.

Thus, in the local theory the kinetic functions $a(\varepsilon_\perp)$ and $b(\varepsilon_\perp)$ (3.33), (3.34) are presented as integral transformations of the kernel (3.24), which was obtained based on the local Boltzmann equation. In the inelastic scattering of CPs, the proposed approach was used to establish a significant difference between the contributions to the kinetic coefficients, which introduce particles from different fractions of the beam, and this, in turn, led to the singularities of the functions $a(\varepsilon_\perp)$ and $b(\varepsilon_\perp)$.

3.7. Transition to the kinetic Boltzmann equation in the theory of rarefied gases

The kinetic equation for a dilute gas of spinless molecules may be obtained by transforming equation (3.15). Transformations in this case are as follows. Since the gas is a single-component system, superscripts g and c, describing the different types of particles, are omitted in all variables. In terms of pairwise interactions of gas molecules, calculations are reduced to a two-body problem and the dynamic variables in (3.16) are replaced by the following

$$\mathbf{q}^g \to \mathbf{q}_1, \, \mathbf{q}_1^c \to \mathbf{q}_2, \, \tilde{\mathbf{Q}}^g \to \mathbf{Q}_1, \, \tilde{\mathbf{Q}}_1^c \to \mathbf{Q}_2, \, \tilde{\mathbf{P}}^g \to \mathbf{P}_1, \, \tilde{\mathbf{P}}_1^c \to \mathbf{P}_2.$$

As for the distribution functions, the functions of the two component system $F_{10}(\tilde{\mathbf{Q}}^g, \tilde{\mathbf{P}}^g, t)$ and $F_{01}(\tilde{\mathbf{Q}}_1^c, \tilde{\mathbf{P}}_1^c)$ are replaced by the one-component system functions $F_1(\mathbf{Q}_1, \mathbf{P}_1, t)$ and $F_1(\mathbf{Q}_2, \mathbf{P}_2, t)$, respectively. As a result, the first order equation (for molecular concentration n_c) for the distribution function for an inhomogeneous rarefied gas takes the form

$$\frac{\partial F_1}{\partial t}(\mathbf{q}_1, \mathbf{p}_1, t) = -\sum_\alpha \frac{p_1^\alpha}{m} \frac{\partial}{\partial q_1^\alpha} F_1(\mathbf{q}_1, \mathbf{p}_1, t) +$$
$$+ \frac{1}{\omega} \int dx_2 \left\{ \Phi\left(\left|\mathbf{q}_1 - \mathbf{q}_2\right|\right); F_1(\mathbf{Q}_1, \mathbf{P}_1, t) F_1(\mathbf{Q}_2, \mathbf{P}_2, t) \right\}. \tag{3.35}$$

Here we have replaced the symbol for n_c by $1/\bar\omega$, where $\bar\omega$ is the volume per one molecule [1]. We emphasize that in the theory of rarefied gases in the collision term it is sufficient to confine considerations to the values of $\sim 1/\bar\omega$, whereas in the theory of channeling, this approximation is not enough and terms $\sim n_c^2$ are considered in (3.16).

The need to consider a term of the order $n_c^2\beta^2$ in the general expression for the collision term (3.16) is due to the fact that the term $\sim n_c^2\beta^2$ contains the pair distribution function of the crystal F_{02}. The presence of this term leads to the fact that at absolute zero temperature $K_{xx} = 0$, which means that at zero temperature scattering of fast particles is realized by lattice vibrations. Thus, the term of the order $n_c^2\beta^2$ in the general expression (3.16) allows us to keep the internal logic of the theory: no temperature – no scattering.

If the spatial inhomogeneity of the gas is small, we can use a rough approximation $Q_1 \approx Q_2 \approx q_1$. In this case, equation (3.35) takes the traditional form of the Boltzmann equation for an inhomogeneous distribution $F_1(\mathbf{q}, \mathbf{p}, t) = \bar{\omega} g(\mathbf{q}, \mathbf{p}, t)$ of a rarefied gas

$$\frac{\partial}{\partial t} g\left(\mathbf{q}_1, \mathbf{p}_1, t\right) = -\sum_\alpha \frac{p_{1\alpha}}{m} \frac{\partial}{\partial q_1^\alpha} g\left(\mathbf{q}_1, \mathbf{p}_1, t\right) +$$

$$+ \iint dp_2\, 2\pi b\ db\ \frac{\left|\mathbf{p}_2 - \mathbf{p}_1\right|}{m} \times \qquad\qquad (3.36)$$

$$\times \left\{ g\left(\mathbf{q}_1, \overset{*}{\mathbf{p}}_1, t\right) g\left(\mathbf{q}_1, \overset{*}{\mathbf{p}}_2, t\right) - g\left(\mathbf{q}_1, \mathbf{p}_1, t\right) g\left(\mathbf{q}_1, \mathbf{p}_2, t\right) \right\}.$$

Here, the vector $\mathbf{q}_2$ uses a cylindrical system instead of the Cartesian coordinate system, and the cylinder axis ξ is assumed to have the form of a straight line parallel to the vector $\mathbf{p}_2 - \mathbf{p}_1$, passing through the point $\mathbf{q}_1$, and the radius of the cylinder is represented by the impact distance. For the adopted geometry of the problem we have

$$\mathbf{P}_1\left(x_1, x_2\right)\big|_{\xi=-\infty} = \mathbf{p}_1, \quad \mathbf{P}_2\left(x_1, x_2\right)\big|_{\xi=-\infty} = \mathbf{p}_2,$$

$$\mathbf{P}_1\left(x_1, x_2\right)\big|_{\xi=+\infty} = \overset{*}{\mathbf{p}}_1, \quad \mathbf{P}_2\left(x_1, x_2\right)\big|_{\xi=+\infty} = \overset{*}{\mathbf{p}}_2,$$

where $\mathbf{p}_1$ and $\mathbf{p}_2$ are the momenta after the collision, $\overset{*}{\mathbf{p}}_1$ and $\overset{*}{\mathbf{p}}_2$ are the momenta before the collision. In addition, the kinetic equation should be supplemented by the normalization of the function $g(\mathbf{q}, \mathbf{p}, t)$, which is defined by (3.17).

The function $g(\mathbf{q}, \mathbf{p}, t)$, defined by (3.36) and the normalization condition, is a function of the distribution of molecules of a rarefied classical gas moving in a macroscopic volume Ω. The difference between the exact kinetic equation of first order (3.35) and the coarse-grained equation (3.36) appears in the transition to the hydrodynamic description of the system. The hydrodynamic equations derived from (3.36) correspond to a macroscopic system with the equation of state of an ideal gas. Therefore, using (3.35), we can obtain an additional correction to the kinetic coefficients (thermal conductivity, viscosity, etc.), as well as amendments to the equation of state of an ideal gas.

Equation (3.36) is easily rewritten in the form applicable in quantum mechanics. It is sufficient to replace $2\pi b \, db$ by $d\sigma$ – the differential scattering cross section, and interpret $g(\mathbf{q}_1, \mathbf{p}_1, t)$ as a one-particle Wigner function. The equation thus obtained is of limited applicability. This is due to the fact that it does not describe the quantum-statistical effects, so that, strictly speaking, it should be used in studies of the kinetics of the non-degenerate quantum gas.

Equations (3.35) and (3.36) were obtained in a fairly formal approach, rather than by direct inference on the basis of a chain of equations. Therefore, before ending this section, we will discuss again the proposals, which form the basis of the theory.

First of all, let us return to the problem of synchronization. We consider a rarefied classical gas with short-range interaction of molecules. It is assumed that the characteristic value of the interaction potential is equal in the order of magnitude to the gas temperature. In this case we can introduce two characteristic parameters of the theory: $\left(a_0^3 / \overline{\omega}\right) \ll 1$, where a_0 is the effective range of the forces; $\tau_{int} = a_0 / \overline{v}$ is the interaction time, where $\overline{v}$ is the average velocity of the molecules. If $t \gg \tau_{int}$, then most of the time the molecules move as free particles with momentum $P_j^{(s)}$ (the superscript s indicates that the system consists of s particles). We introduce the relaxation time of the ratio $\tau_r = \overline{\omega} \, a_0^{-3} \tau_{int}$. During τ_r function F_s is undergoing changes as a result of which it becomes synchronized with the form of the one-particle function $F_1(t)$. If function $F_1(t)$ has changed too much during the time $t > \tau_r$, then, on the one hand, we can speak of a rapid relaxation change of F_s, where $s \geq 2$, on the other hand – considering a rough time scale, we can talk about averaging of changes F_s. (The coarse scale refers to the scale with time intervals greater than τ_r).

However, the time evolution of the correlation function, synchronized with respect to $F_1(t)$, i.e. $F_s = F_s (x_1, x_2, ..., x_s; F_1(t))$, can be described as a slow process, since the evolution of the multiparticle distribution function is determined by the slow change in the one-particle function $F_1(t)$. As for the slow change of $F_1(\mathbf{q}, \mathbf{p}, t)$, the time interval over which the change of this function becomes visible is determined by the change in $\mathbf{q}$ and has a macroscopic scale. Therefore, the lower boundary of this interval of time is always greater than relaxation time τ_r.

Equation (3.35) was obtained using a more general equation (3.15). Of course, the conclusion (3.35) can also be derived in the framework of Bogolyubov' formalism. For this it suffices to choose the solution of the chain of coupled equations [1] for many-particle distribution functions of a real gas, corresponding to the reduced description by the boundary conditions via correlation weakening; splitting all the many-particle distribution functions in the distant past using the one-particle functions. The solutions obtained in the form of expansion with respect to the low density parameter allow to build the Boltzmann kinetic equation. It is important to keep in mind the boundary condition of correlation weakening, and the expansions with respect to the low-density parameter are applicable only to rarefied gas. However, the overall approach to the theory of non-equilibrium processes, based on the abridged description of the non-equilibrium state and the selection of such solutions is typical of many methods of the theory of non-equilibrium processes.

Derivation of the Boltzmann equation, avoiding formalism Bogolyubov' formalism, was discussed in several papers reviewed in [49]. We will mention briefly only those that are methodologically similar to [1].

A system of coupled equations for the distribution of molecules of a rarefied gas was formulated in [49, 50]. In this case a closed equation for F_2 was obtained only in the zeroth approximation, when outside the typical time scale of time–time interaction F_2 was approximated as the product of two functions F_1.

In another paper (Yvon, 1935) attention was given to the equation for the modified binary distribution function F_2, in which the momenta of both particles and their mutual distance appeared as arguments. In addition, a condition was imposed according to which at the distances exceeding the radius of effective action a_0 the function F_2 is approximated by F_1. Finally, the smoothed distribution functions were introduced in [26]. The physical content of this work is very close to [1], but there is also a significant difference: in [26] coarse-grained functions F_1, F_2, F_3,..., were considered, whereas in [1] only unary function F_1 was coarse-grained. Coarsening of F_1 in [1] is defined by its connection with the macroscopic quantity – density.

4

TWO CLASSES OF IRREVERSIBLE PROCESSES IN STATISTICAL THEORY OF CHANNELING

4.1. Phase mixing and dissipative processes Contributions of S- and V-type

It is known that to describe a macroscopic system it is sufficient to determine its internal energy, volume and mass. The study of the system in terms of these variables can be considered as analysis on the thermodynamic level of description. A more detailed study requires an increase in the number of extensive variables.

For example, dividing the volume into a number of cells, an additional variable can be introduced into the theory – momentum of cells. The thermodynamic level of description with the addition of local momenta should be considered as a hydrodynamic description, and, in the general case, the variables in the ensemble are mass density, momentum and energy as a function of the coordinates.

Finally, we can further complicate the analysis, investigating (rather than the total mass of the system) the mass in various states, including examining the distribution of particles in μ-space. This approach is sometimes called the Boltzmann level of description, which is also used in our presentation. Therefore, it is appropriate to introduce the Boltzmann definition of entropy. In the case of a spatially homogeneous system we have

$$S(t) = -\int d\mathbf{p}\, g(\mathbf{p},t) \ln g(\mathbf{p},t).$$

It should be emphasized that the function $g(\mathbf{p}, t)$, included in the entropy equation, satisfies the kinetic Boltzmann equation, so that the evolution of entropy is completely determined by the kinetic equation.

There are a limited number of kinetic equations used to investigate the evolution of entropy. In the vast majority they are equations for spatially homogeneous systems. However, in the motion of fast particles in solids we encounter a completely different situation, since the spatial distribution of atoms in the medium and beam particles is inhomogeneous. In particular, the task of channeling the inhomogeneous system includes a channel wall and a beam of fast particles, and its evolution is attributed to the effective field of the wall and collisions of the channeled particles (CP) with electrons and atoms of the lattice. Here we need a completely different approach than the direct substitution of the kinetic equation in the expression for the time derivative of entropy. It seems to us that in this case it is more efficient to use the local balance equation for the entropy density (3.28), which coincides in form with the local balance equation (3.27) in field theory.

Equation (3.28) describes the evolution of entropy in a non-equilibrium state, and when the system reaches equilibrium, the entropy density takes the equilibrium value $\tilde{S}_{eq}$. Here we must pay attention to two circumstances. Firstly, an irreversible transition to the equilibrium value $\tilde{S}_{eq}$ takes place as a result of collisions, as the entropy source σ_S includes only the contribution term from the kinetic equation. Second, the principle of entropy increase when the system goes to equilibrium is shown in the statement concerning the sign of the entropy source.

However, the time evolution of the inhomogeneous system is caused not only by the effect of collisions. In this case, an important role is also played by the flux of the mean field. An example could be fluxes in inhomogeneous plasma, which makes it necessary to use the Vlasov kinetic equation. Nevertheless, the self-consistent field never contributes to the entropy source. Effects due to the average field only lead to phenomena which are not associated with an increase in entropy.

In the case of a spatially inhomogeneous system, the entropy source σ_S differs from σ_B only by the positively defined factor.

Therefore, considering (3.22), it can be argued that the sign of the entropy source σ_S coincides with the sign of $K(q_x)$ in the case of heat dissipation of the CPs or with sign $\tilde{K}(q_x)$ in the case of electron scattering. Complete calculation of function $K(q_x)$ and $\tilde{K}(q_x)$ is made in Sections 3.5 and 3.6. The results of calculation of the functions

$$\mu_3(q_x) = \frac{K(q_x)}{K_3}, \quad \mu_6(q_x) = \frac{\tilde{K}(q_x)}{K^{(2)}(0)},$$

where K_3 and $K^{(2)}(0)$ are the positively defined coefficients shown in Fig. 3.1. As seen from the figure, $\mu_3 \geq 0$ and $\mu_6 \geq 0$, so that

$$K(q_x) \geq 0 \quad \text{and} \quad \tilde{K}(q_x) \geq 0.$$

Thus, in the channeling of high-energy particles the condition $\sigma_S \geq 0$ is satisfied, and on the example of the channeling theory, we see that the approximation of fast particles to equilibrium is accompanied by an increase in entropy so that the approach to equilibrium is irreversible.

In terms of entropy production all the irreversible processes should be divided into two broad classes. As an example of the phenomenon of the first class we consider the behavior of a gas of non-interacting molecules. Suppose that initially the gas molecules occupy a small fraction of the total volume of the system. In this case, it is clear that the molecules whose velocities are distributed uniformly in all directions, become after a while evenly distributed throughout the volume of the system. The time interval over which the uniform distribution is achieved in the configuration space depends on how the initial state of the gas is defined. For example, for a beam of high-energy particles, whose velocities are concentrated in a narrow cone, this interval is much greater than for the particlea with a large spread of velocity directions.

However, in both cases, the behavior of the gas in space is irreversible. Irreversible phenomena of this type are characterized by the absence of the characteristic time scale and are not accompanied by any increase in entropy of the system. Another example of the irreversible behavior of the particles within the same process is the behavior of inhomogeneous plasma ions, which is described by the Vlasov equation. In general, this class of processes is known as phase mixing processes.

If we assume that the gas molecules collide frequently with each other, then the process of their distribution in the volume of the system will be determined by the specific dynamics of the collisions and macroparameters of the system as gas density and temperature. Now the interaction of molecules fully determines the temporal scale of the kinetic process, i.e. its relaxation time. In accordance with the above concept the process in the conditions of collisions between molecules leads to an increase in entropy, which is a direct proof of its irreversibility. The processes of this class are called dissipative.

In what follows we shall study in detail some specific examples of dissipative processes, but now confine ourselves to the entropy production in the simplest dissipative process. More precisely, we define the rate of growth of entropy in the motion of foreign particles, including Brownian, taking into account friction. We have

$$\dot{S}(t) = \sum_i \xi_i(t) \langle \dot{p}_i \rangle^t = -\beta_1 \frac{1}{m_B} \sum_i \langle p_i \rangle^t \langle \dot{p}_i \rangle^t,$$

where m_B is the foreign particle mass, and the remaining symbols are given in Section 2.1. Given that the average particle velocity is proportional to the friction coefficient ζ

$$\frac{d}{dt} \langle p_i \rangle^t \sim -\zeta,$$

we obtain

$$\dot{S} = -\beta_1 \zeta \sum_i \left(\langle v_i \rangle \right)^2 = 2\beta_1 Q.$$

Here $\langle v_i \rangle$ is the average velocity of i-th particle, and Q is the dissipative function.

Although the relationship of the time derivative of entropy with the dissipative function was derived in a partial case, the correlation holds its shape in both non-equilibrium statistical mechanics and in non-equilibrium statistical thermodynamics. A more concrete expression for the dissipative function can be written. In particular, if ζ is replaced by the polarization coefficient of friction ξ_x^p (2.12), we obtain the expression of Q in terms of the basic parameters of the microscopic theory.

One of the characteristics of non-equilibrium statistical mechanics, which is dealt with in the Chapters 2–4 of this monograph, lies in the fact that this theory can be used at various levels of description (Boltzmann, hydrodynamic, etc.). These levels form a kind of hierarchy, in which the description of elementary processes as they move from one stage to a higher one is enriched with new details.

In conclusion, we would like to introduce a formal concept, which applies to all levels of description of irreversible processes. As an example, consider a stochastic process $n(t)$. For example, the Boltzmann-level description of a rarefied gas $n(t)$ is the density field in the space of the coordinates and momenta, and on another level on another level it may be a vector of extensive variables. Using these definitions, it is possible to write in a purely symbolic form the equation of motion for the average values

$$\frac{\partial \bar{n}}{\partial t} = V(\bar{n}) - S(\bar{n}). \tag{4.1}$$

Individual terms of (4.1) correspond to the dissipative (V) and mechanical (S) contributions. Note that in all real physical situations, V and S do not depend explicitly on time. Dissipative terms V are due to collisions – elementary acts of irreversible processes, whereas the mechanical terms S describe the flow of the phase fluid (mixing). In particular, the entropy source σ_s in (3.28) is completely determined by V-contributions.

In writing (4.1), we completely excluded from consideration the contribution produced by the external fluxes F; the latter can be assumed to be independent of $\bar{n}$. Solutions of (4.1), which do not depend on time, correspond to steady and, in particular, quasi-equilibrium states of the system. Of course it should be borne in mind that here we discuss the mean states of the ensemble. Taking into account the above considerations, it seems important to specify further the importance of dissipation and mechanical inputs in both equilibrium (quasi-equilibrium) states and in the temporal evolution of non-equilibrium systems.

4.2. Phase mixing at shallow depths of penetration (S-contributions)

4.2.1. Phase shifts of the trajectories. The role of anharmonicity of the particles in the process of approach to quasi-equilibrium

To study the motion of particles in shallow penetration it is sufficient to use a simplified model of channeling which does not take into account any multiple scattering on electrons or nuclear scattering. In this model, the impact of environment on the particle is limited to the influence of the continuous potential of the plane (atomic chain). Moreover, using the approach of stochastic theory, we consider the transverse energy $\varepsilon_\perp$ and the distance x of the particle from the center of the channel in the transverse plane as random variables. Then, to study the motion of fast particles it is sufficient to analyze the behavior of $P\left(x|\varepsilon_\perp^0,z\right)$ – the conditional probability density of finding the particle at a distance x at depth z, provided that the entry transverse energy is

$$\varepsilon_\perp^0 = E_0\psi_0^2 + U_{pl}\left(x_0\right).$$

On a purely mechanical level of description it can be assumed that the statistical equilibrium in the subsystem of particles is reached at depth z_0 from which the probability $P\left(x|\varepsilon_\perp^0,z\right)$ for all entry energies is actually independent of z for $z > z_0$. In the classical theory, in the analysis of the transition to equilibrium it is natural to use the trajectory approach, in which the slow spreading of the probability density $P\left(x|\varepsilon_\perp^0, z\right)$ can be extended to as long as the random phase shifts of the trajectories do not mix them up so that the particle distribution becomes uniform in z.

In these circumstances, it is interesting to study the effect of anharmonicity of the planar potential $U_{pl}\left(x\right)$, due to the terms $\sim x^3$, x^4, etc., in the process of phase mixing. At the same time, we do not deny that the additional phase shifts can cause scattering. However, at small penetration depths, where inelastic scattering is still inefficient, the impact of anharmonicity, in other words, S-contributions from (4.1) dominates [51, 52].

It is important to specify the physical situation regarding those particles which entered the channel in the vicinity of the atomic planes (or chains). For simplicity, we assume that the channel is formed by only one pair of atomic chains. Since the atoms of these chains have random thermal displacements in the x-direction, the potential of the channel is a stochastic potential [53]. Of course, the stochastic potential provides a contribution to phase shifting since it determines random changes of the channel width and also of the channel center. However, in this section, these effects will not be considered. The conditional probability density $P(x|\varepsilon_\perp^0, z)$ oscillates with increasing penetration depth z. This is due to the difference of the wavelength of different trajectories. We explain that the difference of the wavelength λ occurs for two reasons: firstly, because of anharmonicity of the planar (or axial) channel and, secondly, because of differences in the impact distances at which the particles enter the channel. This physical situation is shown in Fig. 4.1.

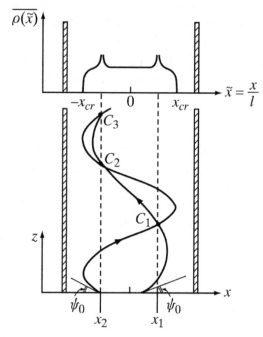

Fig. 4.1. Equilibrium density of channelled particles, relationship of density with particle trajectories.

Along with the conditional probability density we introduce an additional density $\tilde{P}(x;z)$, averaging $P\left(x|\varepsilon_\perp^0,z\right)$ for all possible realizations of the initial transverse energy. We get

$$\tilde{P}(x;z) = \int d\varepsilon_\perp^0 W\left(\varepsilon_\perp^0\right) P\left(x|\varepsilon_\perp^0,z\right). \tag{4.2}$$

From the definition (4.2) it follows that if the conditional probability density varies significantly with increasing z, the same spatial course is characteristic of $\tilde{P}(x;z)$. But the converse is not true. Indeed, $\tilde{P}(x;z)$ can not virtually depend on the depth, while the conditional probability oscillates with increasing z. This strange behavior at first glance of $\tilde{P}(x;z)$ is explained simply enough. It is associated with mixing the components of the integrand in (4.2), more precisely, those components which correspond to different values of the initial transverse energy of the particles. Thanks to the specified mixing, the function $\tilde{P}(x;z)$ can achieve spatial uniformity, even in the absence of scattering particles.

A more detailed analysis shows that in the case where statistical mechanisms are completely excluded, the main contribution to the approach to equilibrium of systems of particles comes from the phase shifts of their trajectories, the latter due to anharmonicity of oscillations of the particles between the walls of a planar (or axial) channel. It should be emphasized that multiple scattering of particles by the electrons and the scattering by thermal vibrations of lattice atoms are regarded as traditional statistical mechanisms.

4.2.2. Evolution of the depth of distribution of channeled particles on the transverse coordinates

Section 4.2.1 dealt with general laws of approach of the CP subsystem approach to statistical equilibrium. We now study the evolution of the spatial distribution in greater detail. To do this, we use the following model representation. Suppose that a spatially homogeneous steady flux of fast particles with energy E_0 is incident on the crystal at an angle $\psi_0 < \psi_{cr}$ to the atomic planes. Furthermore, we assume that the initial angular divergence of the beam particles is much smaller than the critical angle $\delta\psi_0 \ll \psi_{cr}$. Interaction of the fast particles with the medium will be described as a coherent effect of a large number of atoms of the crystallographic plane, i.e. using a continuous non-harmonic potential of the plane $U_{pl}(x)$.

The evolution of the distribution of CPs can be conveniently studied using the Liouville equation for shallow penetration. In this case, the inelastic scattering of fast particles is not considered, and the phase mixing is determined by the gradient of the non-harminic potential, namely by the term

$$C(x) = \frac{1}{2E_0}\frac{\partial}{\partial x}U_{pl}(x).$$

The meaning of $C(x)$ corresponds to that of the S-term in equation (4.1) and, as will become clear from the discussion, does not contribute to the entropy source. If θ is the angle between the velocity of the CP and the atomic plane (y, z), then the Liouville equation for the probability density distribution of the CP to the transverse coordinate x and the transverse velocity v_x, more precisely, the angle $\theta = (v_x/v_z)$, has the form

$$\frac{\partial}{\partial z}\sigma(x, \theta, z) + \theta\frac{\partial}{\partial x}\sigma(x, \theta, z) - C(x)\frac{\partial}{\partial \theta}\sigma(x, \theta, z) = 0. \qquad (4.3)$$

The problem is reduced to the solution of (4.3) with the initial condition

$$\sigma(x, \theta, z)\big|_{z=0, x=0} = \sigma_0(\theta)$$

(X is measured from the middle of the channel) and the boundary condition

$$\sigma(x, \theta, z)\big|_{x=0} = \sigma(x, \theta, z)\big|_{x=2l}.$$

The periodic boundary condition is due to the fact that the continuous potential of the system of parallel atomic planes $U_{pl}(x)$ is periodic in the x-direction with a period equal to the interplanar distance $2l$.

Equation (4.3) with the given initial and boundary conditions was solved numerically, with $U_{pl}(x)$ taken in the form of the Moliere plane potential. The approach proposed in [54] was used for the numerical solution; this approach is a two-level implicit scheme of running counting, stable with respect to initial data. This scheme provides good accuracy of calculation, namely, the second order in all three variables of the distribution of the CPs. The initial distribution is taken in the Gaussian form

$$\sigma_0(\theta) = \left\{2\pi(\delta\psi_0)^2\right\}^{-1/2} \exp\left\{-\frac{(\theta-\theta_0)^2}{2(\delta\psi_0)^2}\right\}. \tag{4.4}$$

Omitting the explicit solution of equation (4.3), we will describe one of its applications. Namely, we consider the evolution of the spatial distribution of particles with increasing z. Finally, we examine an example of the movement of a proton beam with energy $E_0 = 0.5$ MeV in the planar channel of the (100) silicon single crystal. The angular divergence of the beam is $\delta\psi_0 = 0.2\psi_c$, and $\theta_0 = \psi_c/2$. Figure 4.2 shows the graphs of the spatial density of CPs

$$\rho(x,z) = \int d\theta \sigma(x,\theta, z)$$

at different depths of penetration z, Å: 1 – 3600, 2 – 4000, 3 – 4400. The analysis showed that the spatial redistribution of CPs occurs at a depth of ~(0.5 ÷ 0.8) $2l/\psi_{cr}$. (The angular distribution of CPs with the half-width ~ψ_{cr} forms at approximately the same depth). This redistribution is the first stage of an irreversible process – stage of smoothing the probability density due to phase mixing. Further, the spatial density oscillates between the channel walls, and near the

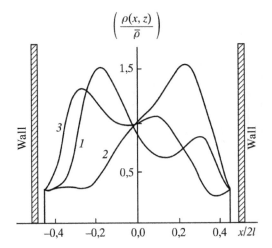

Fig. 4.2. Evolution of spatial distribution with increasing depth of penetration of the channeled particles.

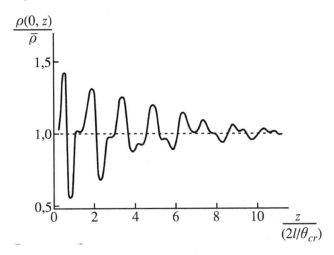

Fig. 4.3. Damping of the oscillations of the spatial distribution of particles in the center of the planar channel with increasing penetration depth.

surface these oscillations are pronounced and are damped at large depths. The oscillations damped with increasing depth of penetration can be seen in Fig. 4.3, which shows the spatial distribution of density in the case of protons moving in the middle of the channel. According to Fig. 4.3, the Liouville equation (4.3) can be used to reveal the fine structure of the evolution of the beam, in particular, the structure of the damping of oscillations in the center of the channel.

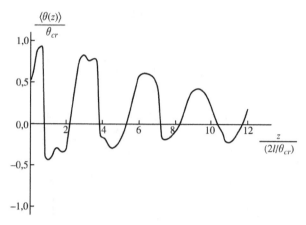

Fig. 4.4. Damping of the oscillations of the mean angle between the plane of the channel and the velocity of a particle with increasing penetration depth.

For particles moving in the middle of the channel, i.e. so-called well-channeled particles, along with the density ρ $(0, z)$, the same procedure canbe used to calculate the dependence of the average values of angle $\langle \theta(z) \rangle$ on the depth z (Fig. 4.4). The mean angle or, in other words, the average transverse velocity also oscillates and decays with increasing z.

However, in the case of $\langle \theta(z) \rangle$ decay is weaker than the rate of reduction of spatial density $\rho(0, z)$. We note that the angular distributions of the beam, determined by this procedure, at different depths fully reproduce the features of the angular distributions observed in the experiment [55]. When the original equation (4.3) does not include inelastic scattering, the only reason for the damping of the oscillations of the density is the anharmonicity of oscillations of the CPs between the channel walls. In other words, the approach of spatial and angular distributions to the equilibrium distribution is determined only by the anharmonicity of the planar potential U_{pl} (x), included in the function C (x) in (4.3).

In general, the attainment of statistical equilibrium in the CP subsystem indicates the simultaneous establishment of equilibrium in configuration and in momentum space. However, in some cases the onset of statistical equilibrium in the momentum space, i.e. equilibrium angular distribution, may not be of any special interest. In particular, when calculating the distribution of the particle flux in the cross section of the channel, we are primarily interested in the total number of particles crossing a surface element, no matter what the direction of the velocity of these particles is.

The question of how the asymmetry of the angular distribution affects the uniform filling of the available area is, of course, not answered. For example, due to blocking [10] in axial channeling, the angular distribution may not be symmetrical. As a result, in some places of the transverse plane the individual elements of the square are shaded and, strictly speaking, the principle of uniform filling can not be used. Yet even in this case, as shown by analysis [56], the asymmetry in the angular distribution has little effect on the uniform filling of the available channel region.

Numerical integration of (1.5) shows that for most of the planar channels the non-harmonic potential $U_{pl}(x)$ in a first approximation can be approximated by a harmonic one. If inelastic scattering is ignored, and the channel capacity is harmonic, then the density $\sigma(x,\theta,z)$, satisfying (4.3), is a periodic function. In this case, in order to obtain the stationary (equilibrium) density $\overline{\rho(x)}$, it is sufficient to average the solution of (4.3) on the wavelength. In addition, during the transition to $\overline{\rho(\tilde{x})}$, where $\tilde{x}$ is measured from the channel center, another factor should be taken into account. The fact is that moving to the regime of directed motion, the particles do come closer to the channel wall than the Thomas–Fermi screening a_{TF} [10]. Therefore, we can introduce another parameter into the theory – the critical deviation of the particle from the center of the channel $d_c = l - a_{TF}$, where l is the half-width of the channel. Adding to the calculation [32] the condition $x < d_c$ on the transverse coordinates of the particle, we obtain

$$\overline{\rho(\tilde{x})} = \frac{1}{\pi}\theta\left(d_c - \tilde{x}\right)\ln\left|\frac{\left(d_c^2 - \chi\right)^{1/2} + \left(d_c^2 - \tilde{x}^2\right)^{1/2}}{\left(\chi - \tilde{x}^2\right)^{1/2}}\right|. \qquad (4.5)$$

Here x_m is determined from the equation $U_{pl}(x_m) = E_0\psi_0^2$, and $m = 1.2$; in addition to this, taking into account the symmetry of the planar channel in relation to the center, we have $x_1 = -x_2$; the value χ included in (4.5) is $\chi = x_1^2 = x_2^2$.

According to (4.5), the stationary spatial density $\bar{\rho}$ has a singularity at points x_1 and x_2, shown in Fig. 4.1. If we take into account any scattering mechanisms, with the exception of coherent, then the singularity of the function (4.5) is replaced by maxima. Note also that the possibility of formation of density peaks (flux peaking effect) near the channel walls was discussed in quantum theory [14], where they were connected with the presence of particles on above-barrier levels. As for the contribution of the random beam $\overline{\rho(\tilde{x})}$, then this contribution is constant and is not taken into account in (4.5).

In accordance with the principles of classical theory, these features $\overline{\rho(\tilde{x})}$ are easy to explain, considering the shape of different particle trajectories. Indeed, the particles included in the channel at the same angle ψ_0, but with different impact parameters, have different amplitude variations in the channel. Therefore, the points of intersection of the trajectories C_1, C_2, C_3 and others define the points x_1 and x_2 in the transverse direction showing a singularity (see Fig. 4.1). Of course, the individual trajectories do not intersect, but the family of trajectories do, so that the peaks have a finite width.

We make two observations. First, as shown by our calculation (see Fig. 4.3), for protons and α-particles with energy $E_0 \sim 01$ MeV, the statistical equilibrium in the subsystem of CPs occurs at a depth of about 10^4 Å. Our calculation is restricted to the limits of applicability of classical theory. However, the same estimate was obtained from a very different position in the quantum theory of channeling [13, 14]. Second, the picture of the evolution of spatial distributions which we obtained is incomplete because it does not take into account the contribution of dissipative processes, which correspond to the V-terms in equation (4.1).

4.2.3. Local and non-local theory of the state density of fast particles

Using the density of states in the phase space, we can study not only the evolution of the distribution of CPs, but also to calculate the density of states in the CPs in interval $(\varepsilon_\perp, \varepsilon_\perp + d\varepsilon_\perp)$, namely, $G(\varepsilon_\perp)$. Of course, the calculation of $G(\varepsilon_\perp)$ is sufficient to consider only the mechanical inputs such as S-terms from (4.1), describing the flow of the liquid phase – mixing.

All the subsequent conclusions are based on Liouville's theorem, which is fundamental to statistical mechanics. The fact is that in statistical mechanics, rather than consider the image of a point of one subsystem at different times, we study a statistical ensemble, i.e. collection of a large number of identical subsystems at the same time. We assume that the motion of phase points representing the state of these subsystems is described by classical mechanics.

Also, keep in mind that the traditional formulation of Liouville's theorem depends on the choice of the phase space as a combination of the configuration space and momentum space. However, for some particular problems a combination of the space coordinates and velocities also allows a simple statement of the theorem. This theorem is modified somewhat in the theory of channeling, which uses a combination of transverse coordinates and transverse velocities, and the distribution of transverse velocity is replaced by angle θ. Then, passing to dimensionless variables

$$\xi = \frac{x}{l}, \zeta = \frac{z}{l}, \ \vartheta = \frac{\alpha}{4}\theta,$$

where $\alpha = 4 (E_0 / K)^{1/2}$ and $K = \pi Z_1 Z_2 e^2 (N2l)/2l$, we study the phase space density in the (ξ, θ, ζ) space.

Now we obtain the explicit form of the phase density, using the Liouville theorem and the method of characteristics. Namely, we find the characteristic features that define such a family of curves in the variables ξ, θ, ζ, in which partial differential equations of type (4.3) are transformed into ordinary differential equations. As such, it is convenient to solve the equations of motion for $\xi = \xi (\zeta)$ and $\theta = \theta(\zeta)$. Then, the rate of change of the phase space density over time along selected trajectories $\xi (\zeta)$ and $\theta (\zeta)$ is zero

$$\frac{d\sigma}{dz} = 0.$$

(For better visibility we give the formulation of the Liouville theorem in the form traditional for the classical theory). If the system is moving in the (ξ, θ, ζ)-space along trajectories with constant density, then for any ζ, in other words, at any depth of penetration of the particles, the phase density can be associated with the initial density σ_0. As a result, the dimensionless density of states of the CPs in the phase space (ξ, θ, ζ) satisfies the relation

$$\sigma\big(\xi(\zeta), \vartheta(\zeta), \zeta\big) = \sigma\big(\xi_i, \vartheta_i, 0\big) \equiv \sigma_0\big(\xi_i, \vartheta_i\big), \qquad (4.6)$$

where $\xi_i = \xi(0), \vartheta_i(\vartheta_0)$.

Next, we should pay attention to the specifics of the channeling effect of high energy particles. Namely, the initial spatial density of the particle beam is uniform, and the initial density in the space of transverse velocity is Gaussian. Moreover, the variance of the Gaussian distribution is fairly small. Also, for simplicity, we assume that the divergence of the beam is vanishingly small and the initial (dimensionless) phase density can be written as

$$\sigma_0 = \left(\xi, \vartheta\right) = \frac{1}{2}\delta\left(\vartheta - \vartheta_0\right).$$ (4.7)

Using (4.6) and (4.7), after a series of transformations [31, 32], we find the desired expression

$$\sigma = \left(\xi, \vartheta, \zeta\right) = \frac{1}{2}\delta\left(\varphi\left(-\zeta, \xi, \vartheta\right) - \vartheta_0\right).$$ (4.8)

$$\frac{\partial\varphi}{\partial\zeta} = -\frac{1}{8}\alpha\frac{1}{E_0}\frac{dU_{pl}}{d\xi},$$

$$\varphi\left(\zeta_i, \xi_i, 0\right) = \vartheta_0.$$ (4.9)

In the absence of inelastic scattering, the phase density $\sigma(\xi, \theta, \zeta)$ is a periodic function of ζ with wavelength λ. As shown in the previous section, as a result of the phase shifts associated with non-harmonic potential U_{pl}, the stationary (equilibrium) value of the phase density σ_{st} is established. In order to determine this stationary value, it is enough to average density (4.8) over the period $\alpha T = (\lambda/l)$. We get

$$\sigma_{st}\left(\xi, \vartheta\right) = \frac{1}{\alpha T}\int_0^{\alpha T}d\zeta\sigma\left(\xi, \vartheta, \zeta\right).$$ (4.10)

In the case of the statistical ensemble, which is discussed in this section, the stationary density (4.10) is a function of the dimensionless transverse energy $e_\perp = \varepsilon_\perp / K$ so the final form is

$$\sigma_{st}\left(\xi, \vartheta\right) = f\left(e_\perp\left(\xi, \vartheta\right)\right).$$

Knowing the stationary phase density (4.10), the general form of the density of states of the CPs in the interval $\left(e_\perp, e_\perp + de_\perp\right)$ can be written as

$$G\left(e_\perp\right) = 2T\left(e_\perp\right)f\left(e_\perp\right).$$ (4.11)

In conclusion, we examine the type of density (4.11) in the case of the non-harmonic potential, including an amendment of the fourth and higher orders

$$U_{pl}\left(\xi\right) = K\gamma\,\mathrm{tg}^2\left(\xi\right).$$ (4.12)

Equation (4.12) includes the numerical factor γ with the order of magnitude equal to 10^{-1}. In the case of the potential (4.12) the dimensionless quantities $e_\perp$ and $T(e_\perp)$ can be written in final form

$$e_\perp = \theta^2 + \gamma \, tg^2 (\xi),$$

$$T = T(e_\perp) = \frac{\pi}{2}(\gamma + e_\perp)^{-1/2}.$$

Now we find the explicit form of function φ, using (4.9) with a planar potential (4.12), and the resulting expression is substituted in (4.8). The phase density is then averaged, using (4.10), and we proceed to the dimensional quantities. As a result, we obtain the desired density of states of CPs

$$G(\varepsilon_\perp) = \frac{1}{2}\gamma K^{1/2}(\gamma K + \varepsilon_\perp)^{-1}(\varepsilon_\perp - \varepsilon_\perp^0)^{-1/2}. \qquad (4.13)$$

Note the following feature of (4.13). In the calculation of $G(\varepsilon_\perp)$ we actually take into account only the flow of the phase liquid in the Γ-space (phase mixing). It is therefore necessary to return back to the division of contributions obtained using (4.1) and it can be argued that the quantity (4.13) includes the contribution of S-terms.

So far we have discussed the density of states in terms of non-local statistical theory. We now consider this physical quantity from the position of the local theory of channeling. To do this, we can again use the Liouville equation (4.3) and appropriate initial and boundary conditions. In addition, as in the non-local theory, we define the initial density of states $\sigma_0(\theta)$ as a function of Gaussian form with a finite angular divergence $\delta\psi_0$ in the channel center. In the absence of inelastic scattering, where the energy of the particles of the beam is preserved, the phase density is periodic in the z-direction and must be averaged according to (4.10). Now, it should be remembered that the density $\sigma(x, \theta)$ obtained in this fashion can be written as $\sigma(x, \theta(x))$, since

$$\theta(x) = \frac{1}{v_0}\left\{\frac{1}{m}(\varepsilon_\perp - U_{pl}(x))\right\}^{1/2}.$$

If the phase density is a function of only one variable – the transverse coordinate x, then according to the formalism of the local theory (Section 2.3.3), the density of states of the particles of the directional beam (per unit volume) with transverse energy in the interval $(\varepsilon_\perp, \varepsilon_\perp + d\varepsilon_\perp)$ should be written in the form of

$$G(\varepsilon_\perp) = \int dx \, \sigma(x, \theta(x)) \frac{1}{\tilde{\varkappa}(\varepsilon_\perp)} \left\{ \frac{1}{m} (\varepsilon_\perp - U_{pl}(x)) \right\}^{-1/2},$$

$$\tilde{\varkappa}(\varepsilon_\perp) = \frac{1}{2m} \int dx \left\{ \frac{1}{m} (\varepsilon_\perp - U_{pl}(x)) \right\}^{-1/2}.$$

(4.14)

Further calculation of the density of states (4.14) can only be performed by numerical integration [57]. The result of numerical integration of the expression of $G(\varepsilon_\perp)$ (4.14), provided that the original function $\sigma(x, \theta, z)$ satisfies the Liouville equation (4.3), is represented by dots in Fig. 4.5 for the case of protons with energies of 0.5 MeV, moving in the channel of (110) single crystal silicon. It was assumed that $\psi_0 = 0.3\psi_{cr}$, divergence angle is 0.1 ψ_{cr}.

Determination of the local theory, as shown in Fig. 4.5, gives a non-monotonic function. The presence of maximum $G(\varepsilon_\perp)$ in the region of low transverse energy is due to the fact that the restriction of a fixed angle of entry into the channel ψ_0 leads to filling (at a depth of about 10^4 Å) of a narrow range of transverse energies near the initial transverse energy $\varepsilon_\perp^0$. However, in Fig. 4.5 there is also a lower maximum. The latter corresponds to particles moving in a quasi-channeling mode, and since the area (or level) of quasi-

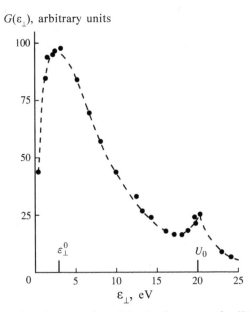

Fig. 4.5. The density of states of particles in the event of splitting of a directional beam into two fractions.

channeling is energetically adjacent to the top of the potential barrier, the second maximum of $G(\varepsilon_\perp)$ corresponds to $\varepsilon_\perp = U_0$.

Recall that the experimental confirmation of the effect of quasi-channeling was discussed in Section 2.5. As in all previous cases, the local theory determines the differences in the contributions to the density of states, which produce the particles moving in the channeling and quasi-channeling modes.

The density of states in the non-local form (4.13) and in the local form (4.14) is equally important for all branches of the theory of channeling, because it is included in the expression for the total energy distribution of the directional beam. In particular, in the quantum version of the theory most of the analytical calculations of the channeling effect are performed using model representations, according to which at the depth of the order of the coherence length statistical quasi-equilibrium is established in the subsystem of the particles of the directional beam.

This is substantiated by the fact that at a specified depth probability amplitudes in the equation of motion for the density matrix decay, but the diagonal matrix elements that characterize the population of individual levels of transverse energy remain unchanged. Rare attempts to formulate the quasi-equilibrium distribution of particles were made until very recently only within the framework of the microcanonical ensemble. At the same time, knowledge of the density of states, as will be shown below, allows the use of the canonical Gibbs ensemble, generalized for non-equilibrium systems, in all problems of the theory of motion of fast particles in crystals.

4.3. The linear theory of reaction

4.3.1. Symptoms of a mechanical disturbance

Consider a classical system with Hamiltonian $H\,(q,\,p)$, independent of time. We will study its response to an external perturbation $H_t^1(q,p)$, corresponding to the interaction with the external field. Assume that the perturbation H_t^1 depends on time

$$H_t^1\left(q,p\right)=-B\left(q,p\right)F\left(t\right),$$

and at $t = -\infty$ is completely absent. Here $B\,(q,\,p)$ is a dynamic value, coupled with the external force $F(t)$. Under these conditions, the distribution function $g(t)$ satisfies the Liouville equation and initial condition: the system is in the statistical equilibrium state with distribution g_0 at $t = -\infty$.

For weakly non-equilibrium states, where $g(t)$ differs only slightly from g_0, the Liouville equation (1.2) and the initial condition determine completely the distribution $g(t)$ at any arbitrary moment of time. If the perturbation H^1_t is weak, the Liouville equation is solved by the iteration method. In particular, we can calculate the mean value of $A(q, p)$ in the linear approximation with respect to H^1_t. We obtain

$$\langle A \rangle = \langle A \rangle_0 + \int_{-\infty}^{\infty} dt' \langle\langle A(t) B(t') \rangle\rangle F(t'), \tag{4.15}$$

$$\langle\langle A(t) B(t') \rangle\rangle = \theta(t - t') \langle \{ A(t), B(t') \} \rangle_0 \tag{4.16}$$

and $\langle ... \rangle_0$ is averaging with respect to distribution g_0, $\{...\}$ are the classical Poisson brackets (1.3).

According to (4.15), the average value of $\langle A \rangle$ is expressed through the retarded double-time Green's function in classical statistical mechanics (4.16). Retarded Green's function determines the average value of $\langle A \rangle$ at time t, provided that the disturbance is included in the time t', for $t > t'$.

Similarly, consider a linear response in the case of a quantum system. The expression for the mean preserves the form of (4.15), but in contrast to (4.16) in quantum statistical mechanics the Green's function takes the form

$$\langle\langle A(t) B(t') \rangle\rangle = \theta(t - t') \frac{1}{i} \langle [A(t), B(t')] \rangle_0, \tag{4.17}$$

where $[A, B]$ denotes the commutator (or anticommutator) and $\langle ... \rangle_0$ – averaging with the equilibrium statistical operator ρ_0. The same definition of Green's function was given in Section 3.5 (formula (3.29)).

Green's function (4.16), (4.17) is a generalization of the time correlation functions and, of course, the impact of perturbation on the average of any dynamical variable can be expressed in terms of the correlation function (Kubo formula [2, 3]). Green's function is applied with the same success to the analysis of systems in a state of statistical equilibrium, and for the study of non-equilibrium processes in which the deviation of the system from equilibrium is small. The Green function method [44, 58] allows us to express the kinetic coefficients in terms of features, including an average over a large

Gibbs ensemble when the occupation numbers of different states are independent and do not need to take into account the constancy of the total number of particles. In addition, the equations of motion for two-time Green's functions can always be supplemented by boundary conditions, which are taken as the spectral theorem [44]. This is a great advantage of the method.

Spectral intensities $I(\omega)$ are useful in calculating the kinetic coefficients. The transfer coefficients, as shown by analysis [44], are proportional to the spectral intensity of the corresponding Green's function.

Spectral representations for Green's functions (we use only the retarded functions) can be easily obtained using the spectral representations for the time correlation functions. Omitting the detailed derivation, we write only the final equation of the spectral theorem

$$G\left(\omega+i\epsilon\right)-G\left(\omega-i\epsilon\right)=-i\left\{\exp\left(\frac{\omega}{\theta_1}\right)-\eta\right\}I(\omega). \qquad (4.18)$$

Here $G(\omega)$ is the Fourier component of Green's function $G(t) = \langle\langle A(t)B\rangle\rangle$; $\epsilon \rightarrow +0$; θ_1 is the modulus of the canonical distribution; $\eta = +1$, if the Green's function is defined on the basis of the commutator, and $\eta = -1$, if on the basis of the anticommutator. Thus, knowing Green's function, we can calculate spectral density using the theorem, and then find an expression for the unknown kinetic coefficient. Below we present the implementation of this program, with specific examples.

4.3.2. Dissipative process (V-contributions) at low velocities of channeled particles

Dissipative processes that occur when the channeled particles (CPs) move in the electron gas have already been discussed in Section 2.2. In this case attention was paid to energy dissipation at high velocities of the CP. In this section, we will continue discussion on this topic, but at low velocities and, importantly, from a completely different position.

The effect of a regular lattice on the fast particles in terms of their correlated collisions with atoms of the plane (y, z), forming the wall of the planar channel, is described by the effective potential $U_{pl}(x)$. In a potential well $U_{pl}(x)$ there are s_0 sub-barrier levels that correspond to states of CPs, performing directional movement along the channel wall in the z-direction. The top of the potential barrier of the well is the level ε_{s_0}. If the plane potential is harmonic, the motion of the CPs is a superposition of the oscillations in the x'-direction around the equilibrium position with frequency Ω and the uniform motion of an equilibrium point with velocity v_0 in the z-direction. Separating the relative transverse movement of the channel wall, we consider the CPs in the moving coordinate system (of course, the thermodynamic parameters do not depend on the choice of the coordinate system). The middle between the two planes forming the channel is denoted by x'_c. Then, the Hamiltonian CP interaction with the electrons (we consider only one-particle excitations) can be written as

$$\tilde{H}_{int}^{(1)} = \sum_{\mathbf{k}} V_k \rho_{\mathbf{k}} \exp\left(-i\,\mathbf{k}\,\tilde{\mathbf{R}}\right),$$

where $\tilde{\mathbf{R}} = \{(x'_c + h(x'_c)), \tilde{\mathbf{r}}\}, h \equiv h(x'_c)$ is the displacement of the CPs, $\tilde{\mathbf{r}}$ is the two-dimensional radius vector of the CPs, V_k is the Coulomb potential.

It would seem that to calculate the dissipative function it is sufficient to consider in $\tilde{H}_{int}^{(1)}$ only the linear term from the shift. However, analysis has shown [59] that restriction to the linear term is only possible if

$$h \ll a_{\mathrm{TF}}.$$

However, the maximum deviation of the CPs can reach a quarter of the channel width $\sim (1/4)a$, where a is the lattice constant, while the screening radius is $a_{\mathrm{TF}} \sim (0.05 \div 0.1)\,a$. This shows that the linear approximation cannot be used.

To overcome this difficulty, we propose the following version of the calculation. We proceed in a coordinate system (x-system), which is fixed relative to the CPs. In this system, the CPs are not involved in the transverse motion relative to the channel wall and the Hamiltonian of the interaction of the CPs with the electrons has the same form as that in the absence of transverse vibrations

$$H_{int}^{(1)} = \sum_{\mathbf{k}} V_k \rho_{\mathbf{k}}\, e^{i\mathbf{k}\mathbf{R}},$$

where $\mathbf{R} = \{x, \tilde{\mathbf{r}}\}$. Coordinate transformation is achieved by a unitary transformation of the interaction operator

$$e^{-iS_q} \tilde{H}_{int}^{(1)} e^{iS_q} = H_{int}^{(1)}, \tag{4.19}$$

where $S_q = -(m_e v_{-q} h)$, provided that the volume of the system is unity, v_{-q} is the Fourier component of the operator of electron velocity.

In the framework of quantum theory, to find the energy lost by the CPs at unit time, it is enough to determine the average rate of change of the unperturbed energy of the CPs

$$n_e Q = \text{Re}\{\text{Sp}(\rho_1(t)\dot{H}_1)\}, \qquad (4.20)$$

where $\dot{H}_1 = -i[H_1, H']$, $H_1 = H_e + H_{int}^{(1)}$ is the complete Hamiltonian of the electrons in the x-coordinate system, H_e is the free-electron Hamiltonian, $H' = \tilde{H}_{int}^{(1)} - H_{int}^{(1)}$ is the interaction Hamiltonian due to the transverse motion of the CPs. The statistical operator of $\rho_1(t) = \rho_e + \Delta\rho(t)$ satisfies

$$\frac{\partial}{\partial t}\rho_1(t) = -i[H_e + \tilde{H}_{int}^{(1)}, \rho_1(t)]. \qquad (4.21)$$

Using the invariance of the trace of the matrix with respect to a unitary transformation, in accordance with (4.19) we make the transition to a moving coordinate system in (4.20). Then the expression $n_e Q$ is expanded in powers of h and we confine considerations to the second-order terms. If the statistical operator is also transformed by a unitary transformation, then

$$\rho_1^S(t) = e^{-iS_q}\rho_1(t)e^{iS_q} = \rho_1 + \delta\rho_1^S(t),$$

where ρ_1 is the quasi-equilibrium distribution of electrons. Using equation (4.21) with (4.19), we find the explicit form $\delta\rho_1^S(t)$.

As a result of all of the above transformations, the dissipative energy of CPs, depending on the momentum transfer $\mathbf{q}$, takes the form

$$n_e Q_{\mathbf{q}} = \text{Re}\left\{\left\langle\left[\left[H_{int}^{(1)}, S_q\right], H_1\right]\right\rangle_1^S + i\left\langle\left[\left[H_{int}^{(1)}, S_q\right], H_1\right], S_q\right]\right\rangle_1^S\right\}, \qquad (4.22)$$

where

$$\langle L\rangle_1 = \text{Sp}(\rho_1 L), \quad \langle L\rangle_1^S = \text{Sp}(\delta\rho_1^S L).$$

Taking into account the formula of the linear theory of reaction (4.15) and knowing the type of $\delta\rho_1^S(t)$, we can express (4.22) by Green's function (4.17). Indeed, after several transformations, we obtain

$$n_e Q_q = -\pi \Omega \operatorname{Im} \left\langle \left\langle f_q \mid f_{-q} \right\rangle \right\rangle_\Omega \left| h \right|^2 . \qquad (4.23)$$

Here $f_q = -i \left[m_e v_q, H_{int}^{(1)} \right]$ is the Fourier component of the fluctuation component of the force applied to the electrons from the scattering center, i.e. by CPs; $\left\langle \left\langle f_q \mid f_{-q} \right\rangle \right\rangle_\Omega$ is the two-time Fourier component of the force–force Green's function.

Next, we integrate the expression $n_e Q_q$ (4.23) over dq with the conservation law taken into account. In writing the law, we should take into account the motion of the equilibrium point of the oscillator at velocity v_0. Then, in the case of low velocities v_0 the energy losses can be written as

$$\left(-\frac{dE}{dt} \right)_V = -\frac{1}{12\pi^2} v_0^2 \frac{1}{n_e} \int dq \, q^4 \operatorname{Im} \left\langle \left\langle f_q \mid f_{-q} \right\rangle \right\rangle_\Omega \frac{1}{\Omega} \left| h \right|^2 . \qquad (4.24)$$

It follows from (4.24) that the energy dissipated by the electronic subsystem is expressed through the imaginary part of the Fourier components of the retarded Green's functions, including the Fourier component of the force f_q. This result is similar to the expression of the theory of motion of foreign particles in an equilibrium medium [27], where the kinetic coefficient – the friction coefficient – is related to the correlation function of the intermolecular forces determined by the microscopic theory.

Description of the dissipative process presented in this section is limited to taking into account the V-contributions in (4.1) and, therefore, so far we have actually taken into account only the dynamic effects of collisions. However, for the analysis of temperature factors at high quasi-temperatures it is also necessary to take into account the kinetic interactions that will lead to a renormalization of the electron energy spectrum. According to the 'hard' zone theorem [60], the perturbation which produces CPs leads to a shift of the parabolic zone as a whole by amount ε_c, where $\varepsilon_c = $ const – the edge of the band.

The final form of energy loss (4.24) expressed through the main parameters of microscopic theory is given in [61].

5

Non-equilibrium statistical thermodynamics as applied to channeling

5.1. General view of the statistical operator for non-equilibrium systems

Formulation of the problem in the linear theory of reactions (Section 4.3) can not always be satisfactory. Indeed, this theory assumes that the selected subsystem is in equilibrium with the thermostat only in the distant past. As for its future evolution, it is considered without the influence of the thermostat taken into account. When $t \rightarrow -\infty$ perturbation H_t^1 takes place and the state of the subsystem evolves according to the Hamiltonian of the subsystem $H(q, p)$ and the operator H_t^1, and the thermostat is eliminated from further consideration.

Let us assume that the external perturbation is not weak, then in the absence of exposure to the thermostat the subsystem may be in a strongly non-equilibrium state and the statistical equilibrium at $t \rightarrow -\infty$ can not be viewed as a zero approximation to its state at any given time. In this case, the zero approximation is the quasi-equilibrium distribution $\rho_l(t)$ [62]. The parameters of the quasi-equilibrium distribution are chosen in such a way that it is quite close to the true non-equilibrium distribution.

The quasi-equilibrium statistical operator $\rho_l(t)$ can be associated with the non-equilibrium statistical operator $\rho(t)$; the latter, unlike quasi-equilibrium statistical operator $\rho_l(t)$, is the integral of the

Liouville equation. Initial attempts to construct the operator $\rho(t)$, using the extremal properties of information entropy failed and the resultant distributions [63, 64] did not describe irreversible processes. The analysis showed that to obtain $\rho(t)$, corresponding to an irreversible process, it is not enough to have only information the maximum entropy values of the thermodynamic parameters at a given time. The functional dependence of the distribution of thermodynamic parameters at any time in the past (the memory effect [62]) must als be taken into account.

Due to the brevity of this section, we refer the reader interested in details of the problem concerned to a fairly complete review [65]. But here we confine ourselves to the most interesting special case of this problem. Assume that the non-equilibrium state is described by a set of average values of a set of operators P_m, where m are integers. Then, the non-equilibrium statistical operator can be written as [65]

$$\rho(t) = \exp\left\{-S(t,0) + \int_{-\infty}^{0} dt' \exp(\epsilon t') \dot{S}(t+t',t')\right\}. \qquad (5.1)$$

Here

$$S(t,0) = \Phi_1 + \sum_m P_m F_m(t),$$

$$\Phi_1 = \ln \mathrm{Sp} \exp\left\{-\sum_m P_m F_m(t)\right\} \qquad (5.2)$$

is the entropy operator and

$$\dot{S}(t,t') = \exp(iHt')\dot{S}(t,0)\exp(-iHt'),$$

$$\dot{S}(t,0) = \frac{\partial S(t,0)}{\partial t} - i\left[S(t,0),H\right]$$

is the operator of entropy production, $F_m(t)$ are the Lagrange multipliers. The operator (5.1) makes it possible to obtain the transport equation and calculate the kinetic coefficients in the case of non-equilibrium systems. However, for the equilibrium case it changes to the Gibbs distribution.

As shown in [44, 65], the non-equilibrium statistical operator (5.1) can be determined by taking the quasi-equilibrium statistical operator as the original expression

$$\rho_l(t) = \exp\left\{-S(t,0)\right\}. \qquad (5.3)$$

It is sufficient to process the operator $\rho_l(t)$ (5.3) by the operation of taking the invariant part. In the non-equilibrium statistical thermodynamics, this operation makes it possible to select the retarded solutions of the Liouville equation. The operator (5.1) satisfies the Liouville equation (1.2), which is complemented by an infinitesimal source on the right side; the source breaks the symmetry of the equations under time reversal. The full procedure of taking the invariant part of the operator will be shown in Section 5.3, in particular, the problem of dechanneling of fast particles.

Parameters $F_m(t)$, included in (5.2), are chosen so that the true average values of P_m were set equal to their quasi-equilibrium average $\langle P_m \rangle_l^t = \mathrm{Sp}(\rho_l P_m)$, i.e.

$$\langle P_m \rangle^t = \mathrm{Sp}\big(\rho(t)P_m\big) = \langle P_m \rangle_l^t. \tag{5.4}$$

With this choice the quasi-equilibrium distribution is very close to the true non-equilibrium distribution, and $F_m(t)$ has the meaning of the conjugate thermodynamic parameters associated with $\langle P_m \rangle^t$.

We assume that the set of variables which determine $\rho(t)$ (5.1), is a set of average values of the operators P_m. Depending on the variable, the selection method of the non-equilibrium statistical operator can be used in both the kinetic stage of the process and in the hydrodynamic stage. In particular, on the hydrodynamic level of description, as a rule, we introduce the energy density of particles and momentum, which depend on spatial variables. Along with the average values of variables, thermodynamic parameters, which are also functions of the coordinates and time, are introduced to the theory.

It should be emphasized that this approach is used at the microscopic level of description, as this was first done in [43]. In addition, it is important to stipulate that the thermodynamic parameters correspond to the thermodynamic parameters of quasi-equilibrium state, which is characterized by the same values of average densities as the non-equilibrium state (the principle of correspondence).

The value of $-\langle \ln \rho(t) \rangle^t$, where $\rho(t)$ given by (5.1), can not be selected as entropy. The fact that $\rho(t)$ satisfies the Liouville equation. If we define the entropy as a chosen expression $-\langle \ln \rho(t) \rangle^t$ then this entropy would have remained constant in an irreversible process, rather than increase. With this in mind, the entropy of a non-equilibrium process should be defined by [65]

$$S(t) = -\langle \ln \rho_l(t) \rangle_l^t.$$ (5.5)

The definition of entropy (5.5) satisfies the thermodynamic equations. Indeed, taking the variational derivative of the Massieu–Planck function Φ_1, we have the thermodynamic relations between the $F_m(t)$ and $\langle P_m \rangle^t$

$$\frac{\delta \Phi_1}{\delta F_m(t)} = -\langle P_m \rangle_l^t = -\langle P_m \rangle^t,$$

therefore,

$$\frac{\delta S(t)}{\delta \langle P_m \rangle} = F_m(t).$$ (5.6)

This confirms the correctness of the definition (5.5).

5.2. Non-equilibrium statistical operator in quantum theory of channeling

5.2.1. Model representations. Thermodynamic parameters

The channeling theory outlined in previous sections, including Lindhard's theory, is based on classical mechanics. We will now study the motion of fast particles from other positions, considering the particles as a quantum subsystem. It is worth recalling that, immediately after discovering the channeling effect, discussion started on the contribution of quantum effects to this new physical phenomenon.

The fact is that, on the one hand, increase of the relative velocity of colliding particles results in violation of the criterion of applicability of classical mechanics since the distance of the smallest particles approach becomes less than the corresponding de Broglie wavelength [66]. On the other hand, the probability of close collisions under channeling conditions is small and, moreover, decreases with increasing particle energy, and the condition of applicability of classical mechanics in the case of channeling of the atoms is satisfied. In contrast, in the channeling of light particles, such as electrons and positrons, the conditions of applicability are violated even at moderately high energies when there is a significant increase in mass due to relativistic corrections. The classical theory is inapplicable to these particles, since the diffraction of electrons and positrons is implemented in the energy range where we can not neglect the quantum corrections.

Thus, the shortcomings of classical approaches to the theory of channeling of charged particles make it necessary, along with the improvement of traditional approaches, to look for new versions of analysis of movement of the CPs. This position was advocated in [67–69]. Moreover, the authors of [70–73] do not exclude the possibility of manifestation of quantum effects even in channeling of the heavier particles, in particular the protons. In this regard it is important to mention the studies [13,14,74] which proposed a quantum-mechanical theory of channeling of atomic particles, which gave new and interesting results, regardless of the actual manifestation of quantum effects. In these papers, the channeling was analysed taking into account the electron and phonon scattering on the basis of the density matrix and the dynamic scattering theory. Some model concepts [13,14] will be used in what follows.

The effect of a regular rigid lattice on the particles in the correlated scattering conditions can be described by the effective periodic potential V_{eff}. If we assume that the particles move parallel to the plane (y, z) in a simple cubic lattice (and in the future we will confine our discussion to planar channeling), the effective potential is one-dimensional $V_{eff}(x)$. In the presence of a potential field $V_{eff}(x)$ the transverse (in x-direction) motion of particles is collectivized, whereas movement in the other two directions is still free. In each well of the potential $V_{eff}(x)$ there is quite a large number of sub-barrier levels. In addition, as shown by analysis of the wave function of particles, in the channeling conditions there are levels of above-barrier states energetically adjacent to the top of the potential barrier.

According to the classical description of the channeling effect at angles less than critical, the angular distribution of particles is divided into two parts: ordered (channeled), and chaotic; in the quantum theory these beam parts correspond to sub-barrier (inside the channel) and above-barrier (outside link) levels. Separation of the angular distribution and the formation of two groups of states occur at a depth of penetration exceeding the depth at which the channeling regime is established (coherence length).

Thus, the transverse motion of particles – the movement in the accompanying frame of reference – can be described by a model potential, formed (in the simplest approximation) by two adjacent rigid planes. Although there is no uncorrelated scattering in a regular rigid lattice, the deflection of the true potential of the medium from $V_{eff}(x)$ leads to single or multiple scattering. As a result, CPs carry out transverse motion between the planes forming the channel, and at the same time experience uncorrelated scattering by electrons and thermal vibrations of the lattice atoms.

As part of model representations, the complete system considered in the problems of channeling can be divided into three weakly interacting subsystems: the thermostat $\{i = 1\}$, which is a lattice or electron gas, the particles in the sub-barrier levels $\{i = 2\}$ and the particles in the above-barrier levels $\{i = 3\}$. Accordingly, the full Hamiltonian of the system $H = H_1 + H_2 + H_3$ consists of the Hamiltonian of the thermostat and the Hamiltonians of the above-barrier and sub-barrier particles

$$H_i = H_0^{(i)} + H_{int}^{(i)} \quad (i = 2,3).$$

Here the operator $H_0^{(2)}$ includes a periodic potential $V_{eff}(x)$, and $H_{int}^{(i)}$ describes the processes of uncorrelated scattering in the i-th subsystem.

We confine ourselves to the spatially homogeneous case. This means that we neglect the influence on the channeling effects of secondary processes, such as transport processes in the configuration space. Then, the non-equilibrium state of the system can be described by a set of average values of a set of operators P_{mi}.

If, for simplicity we exclude the chaotic part of the beam, the rest of the system should be divided further into two parts: $\{i = 1\}$ – the thermostat, which includes the crystal lattice and the electron gas, $\{i = 2\}$ – light atomic particles moving in the channeling regime. The variables P_{mi} are represented by the Hamiltonians of two subsystems $P_{1i} = H_i$, total momenta $P_{2i} = P_i$ and the number of particles $P_{3i} = N_i$. Since the collisions of the CPs with the electrons occur most often, we restrict ourselves to the electron collisions. The Hamiltonian of the CPs can then be written in the form $P_{21} = H_2 = H_0^{(2)} + H_{int}^{(2)} + \tilde{H}_{int}^{(2)}$, where $H_0^{(2)}$ consists of the continuous potential of the atomic chain U_a (or plane) that describes the impact on CPs of the regular lattice in the conditions of the correlated collisions. The total Hamiltonian of the system is $H = H_1 + H_2$. We use the notation: $H_1 = H_e$ – the free-electron Hamiltonian, $H_{int}^{(2)} = W_c^{(2)}$ is the operator of interaction of the CP with one-particle excitations of the electron gas, $\tilde{H}_{int}^{(2)}, = W_p^{(2)}$ is the interaction of the CPs with the plasmons. In this section, the system volume is equal to one.

Assume that the fast particles are initially moving with velocity v_0 along the crystallographic axis (or plane) in the direction z. To eliminate the directional movement of the particles we transfer to the coordinate system that accompanies the particle.

In accordance with the definitions of non-equilibrium statistical thermodynamics (5.2), the operator of entropy and entropy production operator are

$$S(t,0) = \Phi_1 + \sum_{mi} P_{mi} F_{im}(t),$$

$$\dot{S}(t,0) = \sum_{mi} \left\{ \dot{P}_{mi} F_{im}(t) + \left(P_{mi} - \langle P_{mi} \rangle_l^t \right) \dot{F}_{im}(t) \right\}. \tag{5.7}$$

Here

$$\Phi_1 = \ln \text{Sp} \exp \left\{ \sum_{mi} P_{mi} F_{im}(t) \right\}$$

is the Massieu–Planck functional, $\langle P_{mi} \rangle_l^t = \text{Sp} \, (\rho_l P_{mi})$. As for the general form of the non-equilibrium operator $\rho(t)$, then it is still given by (5.1), in which the entropy and entropy production have the form (5.7).

Lagrange multipliers $F_{im}(t)$ in (5.7) are analogous to factors $F_m(t)$ in (5.2), the latter is known to represent the thermodynamic parameters. Unlike the physical situation discussed in Section 5.1, in the channeling problem we study a strongly non-equilibrium system, but in this case, the introduction of single uniform temperature and chemical potential is impossible. Due to the need to divide the parameters relating to different sub-systems, we introduce an additional index i (where $i = 1, 2, 3$), so that factor F_{im} can take three different values. In accordance with (5.4) the correspondence principle is written in the form

$$\langle P_{mi} \rangle^t = \langle P_{mi} \rangle_l^t. \tag{5.8}$$

Then, using (5.8), we vary the entropy (5.7). We get

$$\frac{\delta S(t)}{\delta (P_{mi})} = F_{im}(t). \tag{5.9}$$

More precisely, we obtain the thermodynamic equations, which are direct evidence that $F_{im}(t)$ are the thermodynamic parameters of the i-th subsystem.

Determination of the thermodynamic parameters of the same type as (5.9) was first described in the kinetic theory of gases, and then in the same condition it was formulated in [75–78], and many other works. If the process of transfer of energy, momentum and mass occurs very slowly over time and, moreover, is characterized by small spatial gradients of temperature (quasi-temperature), density and mass flow rate, the equalities (5.9) can be used in both the kinetic and hydrodynamic stages of the process.

5.2.2. Invariant part of the quasi-equilibrium statistical operator. The particle number flux

We begin by considering the relaxation processes that occur in a spatially homogeneous system consisting of weakly interacting subsystems. An example of such processes include: exchange of energy between the two subsystems which is fairly slow because of the small corresponding effective cross-section: the energy exchange between components of a mixture of gases having different temperatures, quasi-chemical reaction in a spatially homogeneous phase (we shall return to this process), etc. One option for such a reaction is an irreversible process that occurs so slowly that the spatially homogeneous state is established in the entire volume. This quasi-chemical reaction is used as a model of dechanneling, of course, the first approximation model.

As mentioned above, to obtain the explicit form of ρ it is necessary to find the invariant part of the operator $\rho_l(t)$ with respect to the motion with the Hamiltonian H. Here $\rho_l(t)$ is the quasi-equilibrium statistical operator (5.3), (5.2), and its invariant part in the future will be denoted by two lines above the operator. Repeating the procedure [44], we

$$\rho = \exp\left\{-\overline{\overline{\Phi}}_1 - \sum_{mi} \overline{\overline{P_{mi} F_{im}(t)}}\right\}, \qquad (5.10)$$

where

$$\overline{\overline{\Phi}}_1 = \ln \exp\left\{-\sum_{mi} \overline{\overline{P_{mi} F_{im}(t)}}\right\},$$

$$\overline{\overline{P_{mi} F_{im}(t)}} = P_{mi} F_{im}(t) - \int_{-\infty}^{0} dt' \exp(\epsilon t') \dot{P}_{mi}(t') F_{im}(t+t') -$$

$$- \int_{-\infty}^{0} dt' \exp(\epsilon t') P_{mi}(t') \dot{F}_{im}(t+t'),$$

$$(5.11)$$

$$P_{mi}(t) = \exp(iHt) P_{mi} \exp(-iHt),$$

$$\dot{P}_{mi} = \frac{1}{i}[P_{mi}, H], \quad \dot{F}_{im}(t) = \frac{dF_{im}(t)}{dt}.$$

Operators $\overline{P_{mi}F_{im}(t)}$ satisfy the Liouville equation in the limit $\epsilon \to +0$. Therefore, the form (5.11) should be considered as an invariant part of the work $P_{mi}F_{im}(t)$ with respect to the temporal evolution of the Hamiltonian H. Expression (5.11) is the quasi-integral of motion equations for $\epsilon \to +0$, and the operator (5.10), built on the basis of (5.11), in turn, is the quasi-integral of the Liouville equation. As a result of taking the invariant part we find the retarded solution of the Liouville equation.

If the system consists of weakly interacting subsystems, the approach to thermodynamic equilibrium is possible in two stages: first – partial equilibrium is established in the subsystems, second - the system comes to a complete balance in general. Two time scales and τ_r and τ_{tot} can be compared with these two stages, and $\tau_r \ll \tau_{tot}$.

In the case of channeling of fast particles, the CPs initially leave the directional motion regime and then, once in the chaotic beam – they are thermalized. (That is why there is no thermalization of the CPs). Given the existence of the two time scales, we can assume that the stationary distribution of particles with quasi-temperature $1/F_{i1}$ and the stationary quasi-chemical potential are established over time τ_r in a macroscopic volume containing a finite number of fast particles (here $i = 2, 3$).

These considerations allow us to greatly simplify the form of the non-equilibrium statistical operator. Indeed, in the limiting steady-state case, from (5.10) and (5.11) we get

$$\rho = Q^{-1} \exp\left\{ -\sum_i (F_{i1}H_i + F_{i2}N_i) - \right.$$

$$-\sum_i (F_{i1} - F_{11}) \int_{-\infty}^{0} dt\, e^{\epsilon t} \dot{H}_i(t) + \tag{5.12}$$

$$\left. +\sum_i (F_{i2} + F_{12}) \int_{-\infty}^{0} dt\, e^{\epsilon t} \dot{N}_i(t) \right\}.$$

Although the statistical operator (5.12) corresponds to the steady state, it can be applied in the case of the non-stationary state, assuming that F_{i1} and F_{i2} change slowly with time.

As a concrete example of the application of (5.12) we calculate the rate constant of dechanneling R. To do this, we average the flow of the number of above-barrier particles $\dot{N}_3$ with the operator ρ (5.12). Retaining the main terms in the resultant expression, we find

$$J_{N_3} = \langle \dot{N}_3 \rangle = L_{\dot{N}_3 \dot{N}_3} \left(F_{32} - F_{22} \right) + L_{\dot{N}_3 \dot{H}_1} \left(\beta_1 - F_{21} \right), \qquad (5.13)$$

where

$$L_{\dot{N}_3 \dot{N}_3} = \int_{-\infty}^{0} dt\, e^{\varepsilon t} \left(\dot{N}_3, \dot{N}_3 (t) \right),$$

$$L_{\dot{N}_3 \dot{H}_1} = \int_{-\infty}^{0} dt\, e^{\varepsilon t} \left(\dot{N}_3, \dot{H}_1 (t) \right).$$

are the Onsager kinetic coefficients. In addition, (5.13) includes the usual designation for quantum correlation functions

$$(B, C) = \int_0^1 d\tau \left\langle B \left(e^{-\tau A} C e^{\tau A} - \langle C \rangle_l \right) \right\rangle_l,$$

$$A = \sum_{mi} P_{mi} F_{im}.$$

We make two remarks concerning (5.13). First, the conclusion can be given for J_{N_3}, taking as a basis the generalized transport equation [15]. The final expression for the flow in this approach coincides with (5.13). Secondly, the flow J_{N_3} due to transitions $\{2\} \rightarrow \{3\}$, which in the classical theory can be interpreted as transitions from the channeled part of the beam in the chaotic one. Since the probability of the reverse transitions $\{3\} \rightarrow \{2\}$ in the experimentally realizable conditions is negligible, the correction for the re-trapping of the particles in the channeling mode is not taken into account.

Concluding the discussion of (5.12) and (5.13), we note once again that both formulas were obtained in the spatially homogeneous model. However, in reality the physical situation is complicated by the fact that the quasi-equilibrium is established in a macroscopically small volume, inside which the fast particle is located at the given moment. The distribution the modulus of which is transverse quasi-temperature is established in this volume, more precisely, within a block of coherent domains [79]. The rest of the volume of the crystal (except for blocks) is in thermodynamic equilibrium with the temperature $1/\beta_1$ which, as will be shown below, is much lower than transverse quasi-temperature.

In this interpretation, there is no contradiction with the general principles of thermodynamics [44], since the non-equilibrium statistical thermodynamics permits the existence of two scales of relaxation time and, most importantly, the short time to establish quasi-equilibrium in a macroscopically small volume. Further development of the thermodynamic theory of dechanneling requires generalization of the expression for the flux in the light of this situation. Indeed, the general scheme of constructing (5.12) and (5.13) can easily be extended to the case in which the individual subsystems are described by the values of $P_{mi}(x)$ and parameters $F_{im}(x, t)$. The entropy of the non-equilibrium state $S(t)$ is the entropy of the equilibrium state in fields $F_{im}(x, t)$, which corresponds to the same density distribution of mechanical quantities $\langle P_{mi}(x) \rangle$ as in the original equilibrium state. This definition of entropy ensures that the thermodynamic equation

$$F_{im}(x,t) = \frac{\delta S(t)}{\delta \langle P_{mi}(x) \rangle},$$
(5.14)

is fulfilled, which confirms the correctness of the thermodynamic parameters.

As the particles move with the directional beam velocity v_0 in the z-direction, the directional movement of the CPs can be prevented by choosing the accompanying coordinate system. The same procedure can be repeated with the particles of the system $\{3\}$, by choosing a coordinate system moving in the z-direction with the average speed of above-barrier particles. As a result of the definition (5.14), in the stationary case

$$F_{21} = \frac{\partial S}{\partial \langle H_2' \rangle}, \quad F_{31} = \frac{\partial S}{\partial \langle H_3' \rangle};$$

$$F_{22} = -F_{21}\mu_2 = \frac{\partial S}{\partial \langle N_2 \rangle}, \quad F_{23} = -F_{31}\mu_3 = \frac{\partial S}{\partial \langle N_3 \rangle}.$$
(5.15)

Here $\langle H_i' \rangle$ is the average energy of the i-th subsystem of particles in the moving coordinate system; μ_2 and μ_3 are the quasi-chemical potentials of the subsystem of sub-barrier and above-barrier particles.

Consider that in the moving coordinate system the fast particles make only transverse motions relative to the channel walls. Therefore $1/F_{21}$ and $1/F_{31}$ are quasi-temperatures connected with the mean energy of transverse motion for the subsystem of the sub-barrier (channeled) particles and the sub-system of the above-barrier particles. In particular, $1/F_{21}$ is the transverse quasi-temperature of the CPs. The importance of introducing such concepts as the transverse temperature was first shown in the problems of the flow of a gas jet into a vacuum, especially when analyzing the molecular velocity perpendicular to the flow lines [80, 81]. In the following we use a a concrete example of directional beams to demonstrate the real physical meaning of transverse quasi-temperature and usefullness of introducing quasi-temperature to the thermodynamics of the fast particles.

5.3. Definitions of the quantum theory of dechanneling

5.3.1. Expressing the rate constant by means of spectral intensity

We confine ourselves to a planar channel. It is assumed that sub-barrier particles are randomly distributed over the wells of the periodic potential $V_{eff}(x)$. The coordinates of the middle of these wells form a set of random variables X. We assume that in any well from the set X there is only one particle and there is no correlation between the coordinates X_i. For example, consider a particle localized in a region that includes X_i. In a potential well X_i there is a finite number of levels (zones with weak dispersion) and, moreover, the level E_3 – the level of the above-barrier states ψ_3 – is energetically adjacent to the top of its barrier (of all the above-barrier energy spectrum, we confine ourselves to the lowest level). The particle in question is either in the above-barrier state ψ_3, or in one of the states of the well.

To describe this alternative is convenient to use variables of external degrees of freedom, to that end, we introduce $c \overset{+}{c}$ and and $d \overset{+}{d}$ – the operators of the occupation number of the state ψ_3 in the well X_i, which have their own values equal to 0 and 1. The state of the particle inside the well X_i will be described using the internal variable which in the harmonic approximation is represented by s – the vibrational quantum number. Naturally, the set s is bounded above by the value $s = s_0$ which corresponds to the top of the potential barrier of the well ω_{s_0} (or the potential of capture of a particle in the channel).

In the framework of the adopted model assumptions, the number of particles in the sub-barrier states is equal to

$$N_2 = N\sum_i \tilde{\rho}(X_i)\overset{+}{d}d, \quad \tilde{\rho}(X_i) = \frac{1}{N_{\perp c}}\sum_{ss'<s_0} W_{ss'}(X_i),$$

where $\tilde{\rho}(X_i)$ is the dimensionless number density of sub-barrier particles (in the variables of internal degrees of freedom); $W_{ss'}$

generally includes a matrix element of the density and the operators acting on the field functions (the explicit form $W_{ss'}$, in the harmonic approximation will be given below); $N_{\perp c}$ is the number of particles in the transverse layer of the lattice. Moreover, the conditions for the variables of the external and internal degrees of freedom are given by

$$\overset{+}{c}c + \overset{+}{d}d = 1, \quad \sum_{ss' \leq s_0} W_{ss'}(X_i) = 1.$$

In the kinetic theory of chemical reactions in a spatially uniform phase [82] the flow of the number of particles, more precisely, the disappearance of a molecule from a potential minimum, is associated with the presence of the absorbing barrier. (Boundary condition of the type of absorbing barrier is introduced in many stochastic models used in the calculation of flows in problems of mathematical physics, see [83]).

The dechanneling rate constant R_e is expressed in terms of the flux dN_2/dt, which is calculated by averaging the operator $\dot{N}_2$. As for the type of $\dot{N}_2$, it can be determined using the following model. Let CPs, randomly distributed in the wells of the effective potential $V_{eff}(x)$, leave these wells with probability $\sim |J_{23}|^2$ and transfer to the above-barrier state, if they reach the top of the potential barrier ω_{s_0}. The transition matrix element J_{23} depends on the static potential of the well. In this model, after several transformations, the dechanneling rate constant can be written as [15]

$$R_e = \frac{\pi}{4}|J_{23}|^2 \int d\omega \, I(-\omega)K(\omega), \qquad (5.16)$$

where $I(-\omega) = I(-\omega, X, X)$ is the spectral intensity of the temporal correlation function, which after averaging over the random distribution of the CPs in the wells of the potential V_{eff} is independent of the coordinate of the filled well X

$$\langle \tilde{\rho}(X)\tilde{\rho}(X,t')\rangle_{(2)} = \int d\omega \, I(\omega)\exp(-i\omega t'). \qquad (5.17)$$

Averaging is performed with the quasi-equilibrium density matrix of the CPs $\langle...\rangle_{(2)} = \mathrm{Sp}(\rho_2 ...), \rho_2 = \mathrm{Sp}_{(1,3)}\rho_I$, and the operators $\tilde{\rho}(X)$ correspond to the number density of the CPs. The function $K(\omega)$ is the spectral intensity of the temporal correlation function, which includes variables of the external degrees of freedom, which (under the assumption that at the depth at which the channeling regime is established the spreading of the initial distribution in space of the transverse energy is small) can be approximated by averaging the initial distribution.

The latter is characterized by a narrow peak near the initial transverse kinetic energy T_0, defined by its angle of entry into the channel, which, in turn, leads to a narrow range of transverse energies ω_0, within which there is localization of the CPs at the time of establishment of the channeling regime. Denoting the center of this interval $\omega_{s1} \approx T_0$, the function $K(\omega)$ can be approximated as $\delta(\omega_{s0} - \omega_{s1} + \omega)$.

We note that the kinetic coefficient (5.16) (5.17) can be expressed through the temporal density–density correlation function $\langle \tilde{\rho}(X)\tilde{\rho}(X,t') \rangle_{(2)}$. The latter is the average product of two densities of sub-barrier particles shifted in time. This form of transport coefficients is quite common and well known from the theory of Green and Kubo [2,3,29,43]. As in these papers, the main result of our conclusion is to establish the connection of a macroscopic quantity – the rate coefficient R_e with the correlation function, which includes the physical quantities of the microscopic theory.

5.3.2. The total Hamiltonian of the system which includes electrons and fast particles

This section explores the process of dechanneling of the particles in the case where the main dechanneling factor is multiple scattering of electrons on the CPs. A model of the electron gas with high effective density, which depends in general on the energy of incident particles, is used. A similar model was successfully applied in [84–86] to calculate the dissipative function of the electron gas when high-energy particles move in the channeling mode.

At CP velocities v_0 of the order of Fermi velocity v_F the electrons of the outermost shells of the target atoms contribute to the effective density. In this case, to describe the motion of CPs in single crystals with BCC and FCC lattices, it is sufficient to take into account the contribution of two distant shells: one of the s-shells and one of the d-shells (for example, for single crystal gold $6s^1$ and $5d^{10}$). Since the ions have a complicated electronic structure, effective density is calculated taking into account the overlapping of shells of the CPs with the shells of target atoms. The resultant expression for the effective density of free electrons is averaged over all possible positions of the CPs in the transverse plane of the channel, which leads to a model of dense gas of free electrons with a uniform spatial distribution.

At higher velocities of the CPs it is important to take into account the additional contribution to effective density of the electrons from the deeper shells of target atoms (in particular, in the case of gold it is the contribution of all shells from $5p^6$ to $4s^2$). The contribution of electrons lying below the outer d-shell results in the effective density, which depends on the distance from the plane of the channel in a direction perpendicular to its direction (planar channeling), which in turn leads to spatial variation of the energy losses inside the channel. While the task in the original formulation is to obtain a general expression for R_e, allowing for passage to the limit of high speeds, the issues of spatial variation of the kinetic coefficient (and effective density) inside the channel are not considered.

According to the classical description of the channeling effect, the angular distribution of fast particles is divided into two parts: directed (channeled), and chaotic. Two parts of the beam in the quantum theory of particles correspond to sub-barrier (inside the channel) and above-barrier (outside the channel) transverse energy levels. Separation of the angular distribution and the formation of two groups of states occurs at a depth of penetration of the order of coherence length L_{coh}, and most of the analytical calculations are carried out using the idea of establishing a quasi-equilibrium at this depth in the space of transverse momenta of the particles.

Indeed, at $L \sim L_{coh}$ probability amplitudes attenuate in the equation of motion for the density matrix of the CPs but the diagonal matrix elements that characterize the population of individual levels of transverse energy particles remain unchanged. Therefore, at a depth $L \sim L_{coh}$ equilibrium distributions can be established for both CPs and for the above-barrier particles, i.e. chaotic particles, since the depth at which internal equilibrium of the chaotic particles is established is smaller than the coherence length. In this section we restrict ourselves to the basic thermodynamic parameter of the CP subsystem – stationary quasi-tempereature $1/F_{21}$.

If we are interested in the behavior of the CPs at not too small depths of penetration, when the details of the initial state are already irrelevant, the number of variables P_{mi}, needed to describe the state of the particles is reduced. If, moreover, the chaotic part of the beam is excluded from consideration and the entire system is split into subsystems: the thermostat $\{i = 1\}$, which is the ideal lattice plus electron gas, and CPs $\{i = 2\}$, then P_{1i} can be represented by the

Hamiltonian of the i-th subsystem, and P_{2i} and P_{3i} by total momentum $\mathcal{P}_i$ and the total number of particles N_i.

Let us specify the types of interaction which include the Hamiltonian of the CPs $P_{12} = H_2$. In the channeling conditions we can consider several scattering mechanisms: scattering of the CPs on the electrons, on the vibrations of the lattice, defects, as well as multiple scattering on the atoms of the surface layer of the crystal. We restrict ourselves to the electron collisions, since they are most frequent in the channeling conditions. As for collisions between particles, in a real situation the concentration of particles in the beam is so small that these collisions do not play a significant role and the corresponding collision integral is not taken into account in the modern theory of channeling when considering transport processes. Interesting in this regard are studies of another class of fast particles – 'hot' conduction electrons in semiconductors.

Studies of cases of low electron density when the frequency of electron–electron collisions v_{ee} is much smaller than the collision frequency of electrons with the lattice v_{el}, and studies of the opposite case $v_{ee} \gg v_{el}$, led to a non-trivial conclusion: it was found that in both schemes of theoretical description the formulas for the quasi-temperature of 'hot' electrons coincide up to numerical factors of the order of unity (this does not exclude, of course, differences in the limits of applicability of formulas). Taking into account these circumstances, in the original version of the theory the interaction among the CPs will not be considered. The full Hamiltonian of the subsystems {1} and {2}, $H = H_1 + H_2$ is then composed of two parts: the Hamiltonian of the free-electron gas $H_1 = H_e$ and the Hamiltonian of the CPs

$$ H_2 = H_0^{(2)} + H_{int}^{(2)}, $$

where $H_0^{(2)}$ consists of the lattice effective potential V_{eff}, in which atoms are fixed in the equilibrium position, and $H_{int}^{(2)}$ describes the processes of uncorrelated scattering of the CPs by the electrons. To understand the internal equilibrium in the subsystem $\{i = 2\}$, as well as to determine the one-particle density matrix of the CPs, it is essential that the uncorrelated scattering is included in Hamiltonian H_2. The formulation of the problem used here is similar to the traditional formulation of the problem of Brownian motion, when we watch the movement of small groups of non-interacting particles that interact only with the environment in thermal equilibrium.

We introduce a concise description of the subsystems by means of two variables: energy and the number of particles – $P_{1i} = H_i$ – Hamiltonian of i-th subsystem, $P_{2i} = N_i$ is the corresponding operator of the number of particles. If we assume that because of the low probability of transition the energy exchange between the individual particles and sub-systems is a slow process, then at a depth $L \approx L_{coh}$ internal equilibrium, which corresponds to the quasiequilibrium statistical operator (5.3), can be established in each of the subsystems. With the steady regime the movement of the CPs is a superposition of the oscillatory transverse motion in the x-direction and free movement in the longitudinal plane of the channel (y, z). The potential of the 'frozen' regular lattice (or more accurately, the potential of a single well from V_{eff}) in the lowest approximation of $h(X)$ – the instantaneous deviation of the CPs from the center X between the two planes forming the channel, is

$$U_{har}^{(0)} = \frac{1}{2} m \Omega^2 h^2 (X).$$ (5.18)

Here Ω^2 is the square of the natural frequency of vibrations of the CPs after averaging the short-wave oscillations, determined by the spatial period of discrete planes. After averaging the steady-state oscillations Ω^2 is equal to the square of frequency for the channel formed by continuous planes. The Hamiltonian of interaction of the CPs with the electrons, considered as individual excitations (model of a gas of free electrons with high effective density) is equal to

$$H_{int}^{(2)} = \sum_{\mathbf{k}} V_k \rho_{\mathbf{k}} \exp(i\mathbf{k}\mathbf{R}), \quad \rho_{\mathbf{k}} = \sum_{\mathbf{k}'} \alpha_{\mathbf{k}'-\mathbf{k}}^{+} \alpha_{\mathbf{k}'},$$

where V_k is the Fourier component of the potential (unscreened) of Coulomb interaction of the CPs with the electrons; $\rho_{\mathbf{k}}$ is the Fourier component of the fluctuations of the electron density in the second quantization; $\alpha_{\mathbf{k}}^{+}$ and $\alpha_{\mathbf{k}}$ are the operators satisfying the Fermi commutation relations; $\mathbf{R} = \{X + h(X), \mathbf{r}\}$ is the radius vector of the CPs, and $\mathbf{r}$ is the corresponding two-dimensional radius in the plane (y, z). Expanding $H_{int}^{(2)}$ into a series in powers of $k_x h(X) < 1$, the first-order term is retained. Then, the processes of inelastic scattering of the CPs on the electrons, which accompany the oscillations of the CPs in a harmonic potential well (5.18), correspond to the operator

$$\delta U = i \sum_{\mathbf{k}} \{k_x h(X)\} V_k \rho_{\mathbf{k}} \tilde{w}_{\mathbf{k}}^{+} \exp(ik_x X) + \text{comp. conjugate}$$ (5.19)

where $\tilde{w}_\mathbf{k} = \exp(i\mathbf{k}\,\mathbf{r})$, $\mathbf{k}$ is the momentum transferred to electrons in the plane (y, z).

In the study of oscillatory motion of the CPs, we restrict ourselves to the collision of particles with one-particle excitations of the electron gas and, furthermore, we will retain only those components of the Coulomb potential which correspond to large momentum transfer $k > k_c$, where $k_c = \omega_p/v_F$, ω_p is the frequency of plasma oscillations. We write the total Hamiltonian of the system in the second quantization. The following expression is used

$$h(X) = \sum_{ss'} \langle s|\delta x|s'\rangle \overset{+}{a}_s\, a_{s'} \exp\left\{-\frac{2\pi}{a} i(s - s')X\right\}, \qquad (5.20)$$

where the matrix element of the displacement $\langle s|\delta x|s'\rangle$ is calculated between the wave functions of the harmonic oscillator. Substituting (5.20) into (5.18) and (5.19) and, moreover, we add to this expression the kinetic energy of the CPs. The result is

$$H = H'_{cp} + H_e + H^{(2)}_{int},$$

$$H'_{cp} = \sum_s \omega_s\, \overset{+}{a}_s\, a_s, \quad H_e = \sum_k T_k \overset{+}{\alpha}_\mathbf{k}\, \alpha_\mathbf{k}, \qquad (5.21)$$

$$H^{(2)}_{int} = \sum_s \sum_{\mathbf{kk'}} \varepsilon_{\mathbf{k-k'}}(s, s+1)\, \overset{+}{a}_s\, a_{s+1}\, \overset{+}{\alpha}_\mathbf{k}\, \alpha_{\mathbf{k'}} + \text{comp. conjugate}$$

where

$$\varepsilon_\mathbf{k}(s, s+1) = U_k(s, s+1)\tilde{w}_\mathbf{k}, \quad \omega_s = \Omega_s,$$

$$U_k(s,\ s+1) = ik_x \langle s|\delta x|s+1\rangle V_k \exp\left\{i\left(\frac{2\pi}{a} + k_x\right)X\right\}.$$

Here $T_k = (k^2/2m) - \mu_1$ with the energy operator H'_{cp} indicates that it is written in the comoving coordinates.

5.4. Fokker–Planck–Kolmogorov finite difference equation

In the case of oscillatory motion of the CPs in planar channels the transitions between a small number of levels of the transverse potential are very likely [87]. If we consider the same transitions also in the case of interaction of the CPs with the electron gas when calculating the kinetic coefficient which depends on Ω, the processes of excitation of plasmons by individual electrons can be neglected.

This is due to the fact that, according to the laws of conservation, the transitions between adjacent levels, with the energy interval between them equal to 10^{-1} eV, cannot be accompanied by processes of generation of plasmons. Therefore, in studies of oscillatory motion of the CPs, we restrict ourselves to the collision of the CPs with one-particle excitations.

Concentrating on the stationary variant of the theory, when $\langle ... \rangle_I = \mathrm{Sp}\,(\rho_I \, ...)$ and ρ_I corresponds to the steady case, we introduce the definition of two-time Green's function

$$\langle\langle A(t)B \rangle\rangle = -i\theta(t)\langle\left[A(t), B \right]_+ \rangle_I, \tag{5.22}$$

constructed on the basis of the anticommutator $[A,B]_+ = AB + BA$. The definition (5.22) shows that the Green functions used in statistical physics differ from the Green functions in quantum field theory [88] mainly by the averaging method. Instead of averaging over the bottom, vacuum state, averaging $\langle ... \rangle_I$ is carried out in the grand canonical Gibbs ensemble. Consequently, Green's function (5.22) depends on both time and temperature. Obviously, when the temperature tends to zero, Green's function (5.22) becomes the usual field Green's function, in which the averaging is performed on the lower energy state. In contrast to the quantum field theory, where the vacuum expectation values are infinite and are discarded as having no physical meaning, in statistical mechanics the averages for the ground state in the thermodynamic limit are given the observed values.

The use of a large Gibbs ensemble (5.22) is convenient primarily because when working with the ensemble it is not necessary to consider additional restrictions on the constancy of the total number of particles, so that the occupation numbers of different states are independent.

In general, Green's function is a two-time function and depends on $t - t'$, but for simplicity we assume that $t' = 0$ and then it takes the form (5.22). multi-temporal Green's functions cannot be used for many problems of statistical physics and considerations must be restricted to two-time functions. The latter are comfortable enough, since they can use a simple spectral relationships that facilitate the solution of the equations for Green's functions. However, they contain a large amount of information about the equilibrium and non-equilibrium properties of systems. From the two-time Green's functions we use the retarded function (5.22), since its Fourier components can be analytically continued into the complex plane of energy.

We represent the number density of the CPs in a form similar to (5.20). We have

$$\tilde{\rho}(X) = \sum_{ss'} \langle s|n|s'\rangle \overset{+}{a}_s a_{s'} \exp\left\{-\frac{2\pi}{a}i(s-s')X\right\},$$

where in the calculation of the density matrix element $\langle s|n|s'\rangle$ integration is performed over the coordinates of one potential well. Taking into account (5.17) and the explicit form of $\tilde{\rho}(X)$, it can be argued that in order to find the spectral intensity $I(\omega)$ it is sufficient to calculate Green's function

$$G_{ss'}(t) = \left\langle\!\left\langle \overset{+}{a}_s(t) a_{s'}(t) B \right\rangle\!\right\rangle.$$

Given the relationship of a discontinuous function $\theta(t)$, included in the definition of $G_{ss'}(t)$, with the δ-function of t, and the equation of motion for the operator $\overset{+}{a}_s(t) a_{s'}(t)$, we can write the equation for Green's function $G_{ss'}(t)$. Naturally, this equation will include two-time Green's functions of higher order than the original. The equation of motion can be derived for them and obtain a chain of coupled equations for Green's functions. The solution of this chain is a rather difficult task, but the system in question contains a small parameter, more precisely, small interaction $H^{(2)}_{int}$ (5.21). Using a small parameter, the chain of equations for $G_{ss'}(t)$ can be uncoupled, i.e. converted into a closed equation for $G_{ss'}(t)$. In particular, the equation constructed using the Hamiltonian (5.21) has the form [16]

$$\left(\frac{d}{dt} - i\omega_{ss'}\right)G_{ss'}(t) + i\delta(t)\left\langle\left[\overset{+}{a}_s a_{s'}, B\right]_+\right\rangle^0_{(2)} =$$

$$= -\left(\lambda_{sp} + \mu_{sp}\right)G_{ss'}(t) + \lambda_{s-1,p}G_{s-1,s'-1}(t) + \qquad (5.23)$$

$$+\mu_{s+1,p}G_{s+1,s'+1}(t).$$

The coefficients of equation (5.23) are

$$\lambda_{sp} = 2\pi \sum_{k_1 k_2 \Delta} \left| U_{k_1 - k_2} (s, s+1) \right|^2 \cdot \left| \left\langle \tilde{p} \middle| \tilde{w}_k \middle| \tilde{p} + \Delta \right\rangle \right|^2 \times$$

$$\times \tilde{n}_{k_1} \left(1 - \tilde{n}_{k_2} \right) \cdot \delta \left(\tilde{\omega}_{k_1 k_2} - \Omega \right),$$

$$\mu_{sp} = 2\pi \sum_{k_1 k_2 \Delta} \left| U_{k_1 - k_2} (s-1, s) \right|^2 \cdot \left| \left\langle \tilde{p} + \Delta \middle| \tilde{w}_k \middle| \tilde{p} \right\rangle \right|^2 \times$$

$$\times \tilde{n}_{k_2} \left(1 - \tilde{n}_{k_1} \right) \cdot \delta \left(\tilde{\omega}_{k_1 k_2} - \Omega \right),$$

(5.24)

where $\tilde{\omega}_{kk'} = T_k - T_{k'}$; T_k is the electron kinetic energy; $p = (p_y, p_z)$ and $\Delta = (\Delta_y, \Delta_z)$ are the two-dimensional momentum of the CPs in the plane (y, z) and the corresponding momentum transfer.

Thus, the desired function $G_{ss'}(t)$ satisfies the Fokker–Planck–Kolmogorov equation, which describes both the coherent nature of the diffraction of the CPs in the regular single crystal (left-hand side of (5.23)) and the inelastic scattering of of the CPs by the electrons, which determine the form of the collision term (right-hand side of (5.23)). The Fokker–Planck–Kolmogorov equation is written in the finite difference form, which is conditioned by the discrete transverse energy in the quantum treatment of the channeling problem.

As noted above, methods based on the Fokker–Planck–Kolmogorov equation have their origin from the theory of Brownian motion and were originally developed by Kirkwood [26] and Green [50] in the assumption about the nature of Markov processes. Therefore, the coefficients (5.24) can be interpreted as the transition probabilities of Markov chains: the probability λ_{sp} corresponds to the transition $s \rightarrow s + 1$, and μ_{sp} to the transition $s \rightarrow s - 1$.

5.5. The statistical coefficient of dechanneling rate

5.5.1. The kinetic coefficient in quantum theory

Equation (5.23) has an analytic solution for a class of higher transcendental functions. Using the method of solving equations in the finite difference form (5.23) [87], we find

$$G_{ss'm'm}(E) = \frac{1}{2\pi} \delta_{sm} \delta_{s'm'} \sum_{s''} B_{s''1} F_1(-s'', 1; r') \times$$

$$\times \left\{ E + \omega_{ss'} - i \operatorname{Im} M_{s''}(E) \right\}^{-1}.$$

(5.25)

Here $_1F_1(a, \delta; x)$ is a confluent hypergeometric function; $\omega_{ss'} = \omega_s - \omega_{s'}$; $E = \omega \pm i\epsilon$, $\epsilon \to +0$, and the imaginary part of the mass operator [58] is

$$\operatorname{Im} M_{s'}(\omega \pm i\epsilon) = \gamma_{s'} = \mp \frac{1}{\tau_e} \frac{\omega_{s'}}{\omega}.$$

Coefficient $B_{s''}$ includes the value of $r' = s + \Delta s$ [16], and the link of Δs with s is defined by the equation

$$\Delta_s = \xi\{s + (s+1)\tilde{\sigma}_s\},$$

where

$$\tilde{\sigma}_s = {}_1F_1(s+1,1;s) / {}_1F_1(s,1;s).$$

Coefficient $B_{s''}$ from (5.25) satisfies the transcendental equation

$$\sum_{s''} B_{s''} \, {}_1F_1(-s'',1;r') = 2n_s, \tag{5.26}$$

where $n_s = \left\langle a_s^+ a_s \right\rangle_{(2)}^0$ is the quasie-quilibrium distribution of the CPs, normalized to unity.

Using the explicit form of the interaction Hamiltonian $H_{int}^{(2)}$ (5.21), an expression is derived for the energy relaxation time. We find

$$\frac{1}{\tau_e} = 2 \left(\frac{2l}{L_z}\right)^2 \sum_{k_1' k_2'} P_{k_1' k_2'}(\Omega) I_{\tilde{k}_1' \tilde{k}_2'},$$

$$P_{k_1' k_2'}(\Omega) = \frac{\pi}{m_e} \beta_1 \left| V_{k_1' - k_2'} \right|^2 \left\{ (k_1')_x - (k_2')_x \right\}^2 \times$$

$$\times \tilde{n}_{k_1'} \left(1 - \tilde{n}_{k_2'}\right) \cdot \delta\left(\tilde{\omega}_{k_1' k_2'} - \Omega\right), \tag{5.27}$$

$$I_{\tilde{k}_1' \tilde{k}_2'} = \frac{2\pi}{L_y} \sum_{\Delta y} \delta\left(k_y' - \Delta_y\right) \cdot \delta\left(\zeta_{\tilde{k}_1'} - \zeta_{\tilde{k}_2'}\right).$$

Formulas (5.25)–(5.27) completely determine the Fourier component of Green's function $G_{ss'}(t)$ and hence the spectral density $I(\omega)$ (4.18). Since $K(\omega)$ is the spectral intensity of the correlation function with variable external degrees of freedom equal to $\delta(\omega_{s_0} - \varepsilon_{in} + \omega)$, then using (5.16) we easily obtain the dechanneling rate coefficient R_e. Indeed, with the accuracy up to second order in the interaction potential V_k, we have

$$R_e = \frac{1}{4}\left(\frac{1}{\varepsilon_{s_0}}\right)^2 \frac{1}{\exp\left(F_{21}\varepsilon_{s_0}\right)+1}\left|J_{23}\right|^2 \times$$

$$\times \sum_{s''}\sum_{s \langle s0} B_{s'} \frac{1}{\tau_e} \cdot \frac{s''}{s_0}\, {}_1F_1\left(s'',1;r'\right),$$

(5.28)

where ε_{s_0} is the energy potential barrier (or the transverse energy of particle capture in the channeling mode), measured from the energy of entry into the channel $\varepsilon_{in} = E_0\psi_0^2$. Returning to the classification which determines the equation (4.1), it can be argued that the rate constant (5.28) takes into account the contribution of V-members.

In this chapter, we aim to study the changing nature of dechanneling due to reduction of CP energy $E_0 = (p^2/2m)$, and explore the process of dechanneling due to multiple scattering of the CPs, which leads to the population of the highest energy levels of transverse motion up to the level of the potential barrier ω_{s_0}. Therefore, in deriving $1/\tau_e$ we have neglected in the matrix elements the change in longitudinal momentum of the CPs. As shown in [52, 89], the approach of the constant longitudinal velocity of the CPs $v_0 = (p/m)$ is sufficient for most applications of the theory of the transverse oscillatory motion of the CPs in planar channels, in particular, the geometrical aspects of the energy loss spectra of the CPs are discussed in this approximation. Calculation of the correction to the frequency due to the change in velocity of the CPs v showed that the relative magnitude of this correction is negligible (less than 1%).

The problem of finding the rate constants of diffusion de-channeling can be solved in two ways: 1) determine the probability of a transition that was incorporated into the formula for the diffusion coefficient [13], 2) develop and solve the equation for the corresponding transition probabilities for time t [90] or, equivalently, write and solve equations for the time correlation functions, $G_{ss'm'm}(t)$, used to determine R_e.

In the first case, the initial rate constant, with the accuracy equal to a constant factor, is equal to $Q(0) \sim 1/\tau_0$, where $1/\tau_0$ is expressed in terms of quantum-mechanical transition probability $W\left(q_\perp, q_\perp'\right)$. In the second case, the rate constant (5.28), in addition to the characteristics of the potential well, involves the thermodynamic probability of a transition to the level ω_{s_0} (the latter corresponds to the top of the potential barrier), which is based on the consideration of the relaxation process with transition probabilities λ_{sp}, μ_{sp}. In the special case when damping is small, $\gamma_s^2 \ll \Omega^2$, the rate constant (5.28) is inversely proportional to energy relaxation time τ_e (5.27), similar to that in [91] $Q(0) \sim 1/\tau_0$. Comparison of τ_e with τ_0 shows that both values are an integral transform of the effective scattering cross section of the CPs.

We note one more circumstance. The solution of equation (5.23) showed that the attenuation of the correlation of the number density of the CPs is determined by superposition of exponential functions with a number of relaxation constants, and consequently, the behavior of $G_{ss'm'm}(t)$ for large t corresponds to the asymptotic form of correlation functions in the stochastic case. The Fourier component of the function (5.25) contains the relaxation constant in the denominator, because

$$\operatorname{Im} M_s \left(\omega_{s_0} \pm i\epsilon \right) = \gamma_s = \mp \frac{1}{\tau_e} \frac{\omega_s}{\omega_{s_0}}.$$

5.5.2. Semi-classical approximation

The expression for the rate coefficient (5.28), obtained in the previous section, is rather complicated. In order to express R_e by the basic parameters of microscopic theory, one must first perform the summation over $\mathbf{k}_1'$, $\mathbf{k}_2'$ in (5.27), as well as the summation over s and s'' in (5.28). These summations can be performed only in the semi-classical approximation.

Let's start with the transformation (5.27). Namely, the summation is carried out with respect to electron variables $\mathbf{k}_1'$ and $\mathbf{k}_2'$ in the following assumptions:

– We neglect thermal 'smearing' of the distribution functions of electrons, treating the electron gas in the ground state.

– We use the simplest model of the low-index planar channel, considering the atomic plane as a set of atomic chains, separated by a distance equal to d_s. We assume that the channeled particle is moving along one of them (in the direction of z) so that the angle between the direction of its motion and the chain does not exceed ψ_{c1} (1.9). In this case, a single chain stands out in the plane so that it is sufficient to study the motion of channeled particles between the two atomic chains with the oscillation frequency Ω (between chains) and the amplitude u_x in the transverse direction.

– We take into account collisions of the CP with one-particle excitations of electrons due to the close part of the Coulomb interaction are accompanied by large momentum transfers

$\Delta_y \gg (\Omega / v_F)$. In summation, we take into account the diagram of ion – electron collisions in the momentum space, corresponding to the case of high velocities of the CPs $v_0 > v_F$ [92].

– Since the uniform spatial distribution is considered for the valence electrons, we do not take into account their differences with the conduction electrons and introduce uniform density $n_V = Z_V n_c$.

The relaxation constant $1/\tau_e$, obtained within these approximations, is expressed (up to a factor) in the form of an integral transform

$$\frac{1}{\tau_e} \sim \Omega l^2 \frac{m_e}{q_{TF}^2} \frac{1}{k_F} \int d\sigma(\Delta_y) \Delta_y \pi_1(\Delta_y). \qquad (5.29)$$

Here

$$d\sigma(\Delta_y) = \frac{2\pi Z_1^2 e^4}{m_e v_0} n_c Z_V \frac{dQ_y}{Q_y^2},$$

is the scattering cross section; $Q_y = (\Delta_y^2 / 2me)$ is the energy transferred in the limit of high particle velocity; q_{TF} is the wave vector in the Thomas – Fermi model

$$\pi_1(\Delta_y) = \frac{q_{TF}^2}{4\pi} \frac{k_F}{2\Delta_y} \left\{ \frac{\Delta_y}{k_F} + \left[1 - \left(\frac{\Delta_y}{2k_F} \right)^2 \right] \ln \left| \frac{\Delta_y + 2k_F}{\Delta_y - 2k_F} \right| \right\}.$$

Note that the expression $\pi_1(\Delta_y)$ includes the contribution of the first order to the polarization operator $\pi(k, \omega)$ at $\omega = 0$ because we have restricted ourselves to the 'quasi-static' limit $\Omega \ll v_F \Delta_y$ and have not kept fully the space – time correlation, which contains the density–density correlation function (5.17). At the same time, it is not attempted to take into account 'wake' polarization [93], occurring in an electron gas with uniform motion of fast charged particles in the z-direction. Thus, one could argue that the description of the relaxation constant on the basis of (5.29) is directly connected with the relations of the theory of irreversible processes [94], where the Fourier component of the pair correlation function, which includes the number density of particles, is associated with the inelastic scattering cross section and the effects of polarization of the medium.

The final expression for $1/\tau_e$ is substituted into (5.28). Further, taking into account the singularity of the semi-classical approximation, we replace in (5.28) ω_s by continuous variable ε (and, therefore, ε_{s_0} by ε_t).

As for the 'dumb' index s'', the transition to the quasi-classical limit does not require modification of the index. However, in order to ensure the rapid convergence of series in s'', we used the technique which was successfully used (see, for example [82]) to solve boundary value problems with an absorbing barrier. Namely, we replace s'' a discrete non-integer value

$$\mu_r(\zeta) = j_{or}\left\{\frac{\zeta}{(2j_{01})^2}\right\},$$

which is the root of the transcendental equation and depends on the dimensionless parameter $\zeta \gg 1$. Here j_{or} is the r-th root of the Bessel function. In the problems of the disappearance of particles from the potential minimum (or exit from the channel) the dimensionless parameter ζ is represented by the ratio of the potential barrier height to the characteristic energy of the particles, more precisely, to Ω.

The transition matrix element is written in the semi-classical approximation

$$J_{23} = 2(2\pi\Omega)W(\varepsilon_t),$$

where $W(\varepsilon_t)$ is the coefficient of passage through the parabolic potential barrier. In the case of the parabolic barrier $W(\varepsilon_t)$ is approximated by a constant [95].

All the details of calculating R_e (5.28) are given in [17], and we refer the reader to the cited paper. But here we confine ourselves to the representation of the final form of the rate coefficient R_e for efficiently channeled particles. If the result of the phenomenological theory of dechanneling are generalized in the framework of non-equilibrium statistical thermodynamics, the rate coefficient can be written as

$$R_e = C_1 k(F_{21})\frac{1}{E_0 \psi_{cr}^2}\frac{m_e}{m}\left(-\frac{dE}{dt}\right)_V. \qquad (5.30)$$

In the phenomenological theory [96,97], which is based on the relations of classical mechanics, the dimensionless rate coefficient is equal to $k(F_{21}) = 1$, and the numerical coefficient C_1 is assumed to be 0.63. In our derivation of R_e we used the method of the non-equilibrium statistical operator, so the rate constant is a function of the transverse quasi-temperature of the CPs (5.15)

$$k\left(F_{21}\right) = \left\{\exp\left(F_{21}\varepsilon_t\right) + 1\right\}^{-1} \cdot \exp\left(-\mu_1\left(\zeta\right)\right). \qquad (5.31)$$

The energy losses in the model, which considers the CPs as oscillators with frequency Ω and amplitude u_x in the x-direction, are expressed in terms of basic parameters of the microscopic theory and the characteristics of the oscillator

$$\left(-\frac{dE}{dt}\right)_V = \frac{4\pi Z_1^2 e^4}{m_e v_0} n_c Z_V L_e \frac{1}{\omega_p \tau_1}, \qquad (5.32)$$

where

$$L_e = \ln\left|\frac{2m_e v_0 v_F}{\omega_p}\right|, \quad \frac{1}{\tau_1} = \Omega \frac{|u_x|^2}{a_{TF}^2}.$$

We have formulated the statistical theory of channeling using the method of statistical ensembles for non-equilibrium systems, generalizing the usual Gibbs method of ensembles. This transfer of Gibbs' ideas to non-equilibrium statistical mechanics is reflected in the formulation of the non-equilibrium statistical operator (5.12) and the subsequent introduction of the Green function (5.22).

It should be noted that the theory also includes the thermodynamic parameters F_{i1} and F_{i2}. Therefore, we should discuss in greater detail the development of non-equilibrium statistical thermodynamics of the channeling effect. In this case this is possible if we are interested in the behavior of the system for not too small time scales, when the details of the initial states of the system are already irrelevant and the number of variables needed to describe it is reduced..

Speed coefficient. The studies of the rate of release of particles from the channeling mode are mostly based on methods that do not allow analytical expressions to be derived for the kinetic coefficients. The calculation is limited to numerical integration of the equations (see numerical analysis of $D(E_\perp)$ in [98,99]). As for the energy losses, the calculation of these losses (see, e.g. [46,100]) is reduced to numerical integration of the braking cross-section determined by the quantum mechanics methods. Another direction of channeling theory, based on the methods of non-equilibrium statistical thermodynamics, is suggested in [15,16]. Indeed, averaging $\langle...\rangle_{(2)}$ (5.17) carried out with the quasi-equilibrium distribution ρ_l, leads to the final expression (5.30) of the additional function $k(F_{21})$, which depends on the transverse quasi-temperature of the CP subsystem. At $F_{21}\varepsilon_t \ll 1$ this additional function can be approximated by the expression

$$k(F_{21}) = \chi \exp(-F_{21}\varepsilon_t), \tag{5.33}$$

where $\chi = 1/2 \exp(-\varepsilon_t/10\Omega)$ is the pre-exponential factor which does not depend on the transverse quasi-temperature.

The rate constant (5.33) is written in the same form as the diffusion rate constant in configuration space [82], known from Arrhenius kinetics. However, the diffusion problems (and reactions) concern a system close to thermodynamic equilibrium, and the temperature of the subsystem of migrating atoms is close to the temperature of the thermostat, the problems of channeling are characterized by a substantial 'gap' in the thermodynamic parameters of the CP subsystem and the corresponding parameters of the thermostat. In this sense, the rate constant (5.33) has no counterpart in the equilibrium gas-phase kinetics [101].

The dependence of losses on the frequency and amplitude of oscillations of the CPs. The total energy losses of the CPs, in addition to $(-dE/dt)_V$ (5.32), include an additional term $(-dE/dt)_{pl}$, which does not depend on the orientation of the particle beam relative to the crystallographic planes, so that

$$\left(-\frac{dE}{dz}\right) = \frac{1}{v_0}\left\{\left(-\frac{dE}{dt}\right)_{pl} + \left(-\frac{dE}{dt}\right)_V\right\}. \tag{5.34}$$

If the losses $(-dE/dt)_V$ are due to multiple scattering of the CPs on the individual excitations of the electron gas, the losses $(-dE/dt)_{pl}$ are associated with the excitation of plasmons in the scattering of the CPs. The losses $(-dE/dt)_{pl}$ can be expressed by Green's function, which includes the Fourier component of the electron density ρ_q,

$$\left(-\frac{dE}{dt}\right)_{pl} = -2\pi\sum_q \int d\omega\, \omega |V(q)|^2 \, \mathrm{Im}\left\langle\langle\rho_q | \rho_q^+\rangle\rangle^{(0)}_\omega \delta(\omega - qv_0)$$

and, of course, they do not include parameters such as Ω and u_x. Therefore, according to (5.32) and (5.34), for the electronic losses

$$\left(-\frac{dE}{dz}\right) = \alpha + \beta\Omega, \tag{5.35}$$

where the quantities α and β are independent of frequency. Empirically, the frequency dependence of the energy losses of the form (5.35) was obtained in [102] where the channeling of oxygen ions in the planar channels (111) (100) of a gold crystal was studied. However, no theoretical explanation was proposed. Therefore, the overall result of the model used by us and the semi-classical approximation (5.35) are of undoubted interest.

As for the dependence of the energy losses on amplitude, it is completely determined by the harmonic approximation, which provides a 'frozen' potential of the plane (5.18). Indeed, in this approximation, the deviation of the CPs from the middle of the channel is equal to $h(z) = u_x \cos(\Omega t)$, where $z = v_0 t$. After completing the expansion of the energy losses into a series with respect to $h(z)$, we have, with the accuracy up to the second order terms

$$\left(-\frac{dE}{dz}\right) = A(E) + B(E)h^2(z).$$

If we average the second term over the oscillation period, we find

$$\left\langle B(E)h^2(z)\right\rangle_{cy} \sim |u_x|^2,$$

so that the amplitude dependence is consistent with the result (5.32) and (4.24). Note also that formula (5.32) applies only to the well-channeled particles.

The generalized isotopic effect. As we explore the channeling of fast charged particles, it is necessary to consider the dependence of kinetic coefficients not only on the mass of particles, but also on their atomic number. This is a generalization of the traditional isotopic effect. According to (5.30)–(5.32), the dependence of $(1/v_0) R_e$ on the mass of the CPs (at fixed energy) is completely determined by the frequency dependence

$$\frac{1}{v_0} R_e \sim \frac{1}{E} Z_1 \Omega \sim \frac{1}{E} Z_1^{3/2} m^{-1/2}.$$

It follows that the ratio of the rate factor of dechanneling of deuterium to the hydrogen coefficient (for the same Z_1, E) is equal to $\sqrt{1/2}$; for comparison, the value of this ratio obtained in [96] for the channel of (100) germanium was 0.8. The energy losses (5.32) also depend on the mass and atomic number of particles

$$\frac{1}{v_0}\left(-\frac{dE}{dt}\right)_V \sim \frac{1}{v_0^2} Z_1^{5/2} m^{-1/2}.$$

This result, related to the fast particles, is significantly different from the conclusion of the theory of braking of slow particles [61], according to which the electronic losses do not depend on the particle masses and at the same Z_1 are defined only by their velocity v_0.

5.6. The rate of release of particles from the channeling regime, taking into account the spatial separation of directional beam

5.6.1. Resonant transition of particles to the quasi-channeled fraction and relaxation frequency

Now let us consider a more complex version of the dechanneling effect when we take into account the spatial separation of the directional beam into channeled and quasi-channeled parts (the trajectories a_1 and c_1 in Fig. 2.4).

In the quantum model of channeling the particle motion of the directional beam in the channel plane is free and the corresponding energy is continuous. At the same time, the energy of transverse motion is discrete. To give an idea of the quantum states of particles, it is sufficient to enter two modelling potentials:

$$U_2(x') = \frac{m}{2}(\Omega Q(x'))^2,$$

$$U_3(x'') = \frac{m}{2}(\Omega Q(x''))^2 + \omega_{s_0}.$$

Here, $U_2(x')$ and $U_3(x'')$ are respectively the potentials of the CPs and of a particle moving in the quasi-channeling mode (Fig. 5.1), $Q(x')$ is the transverse displacement from the equilibrium position x',

$$\Omega = \frac{\pi}{l}\left(\frac{\omega_{s_0}}{m}\right)^{1/2}$$

is the frequency of transverse vibrations; ω_{s_0} is the potential of capture of a particle in the channeling mode which corresponds to the classical theory of barriers U_0.

If the minimum $U_2(x')$ is in the middle of the channel x', then the minimum $U_3(x'')$ is in the immediate vicinity of the atomic planes forming the channel wall. In the planar potential $U_2(x')$ there is a large number of energy levels $\omega_s \stackrel{+}{=} \Omega_s$, where s is the vibrational quantum number. To describe the state of the CPs, it suffices to define the occupation numbers of quantum states n_s^0. Numbers n_s^0 are the eigenvalues of the operator $a_s^+ a_s$, where the operators a_s and a_s^+ satisfy the commutation Fermi relations.

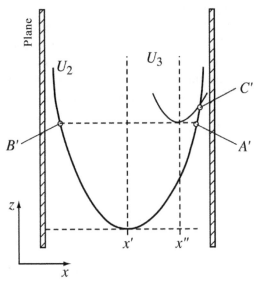

Fig. 5.1. Model potentials of the particles in the event of the directional beam into two fractions.

As for the states of the quasi-channeled particles, from all states in the well $U_3(x'')$ we restrict ourselves to styding its ground state with energy $E_3^{(0)}$. The wave function of the ground state has a maximum near the channel wall [14]. For a more detailed description of the excited states of the hole $U_3(x'')$ we can introduce finite width zones, and with increasing number of the zone the motion of a particle becomes very similar to free motion, i.e. to the chaotic motion of a particle of the chaotic part of the beam.

Thus, the system under consideration is divided into three parts: 1) thermostat $\{i = 1\}$, including the lattice and the electron gas, 2) the chaotic part of the beam, and 3) the directed particle beam, which has two components, namely, CPs $\{i = 2\}$, which are in s states in multilevel wells $U_2(x')$, and quasi-channeled particles $\{i = 3\}$ in the ground state potential wells $U_3(x'')$.

Since the quasi-channeled particles quickly travel to the chaotic part of the beam, the states of the subsystem $\{3\}$ are regarded as short-lived ones of the two components of the directional beam. Therefore, in the study of the dechanneling, i.e. release of the CPs from channeling regime, it suffices to consider the rate of transition $\{2\} \rightarrow \{3\}$. In general, the speed of this transition was studied in Section 5.5.1.

Consider a resonant transition (resonance in respect of transverse energy). If we have the energy diagram (see Fig. 5.1), the resonant transition of a particle from the state {2} to state {3} occurs only if as a result of random walks between the levels of the transverse energy of the well U_2 the particle reaches the level ω_{s_0}, which is energetically adjacent to $E_0^{(3)}$. The transition {2}→{3} is either through point A' or through point B'. In contrast to the quantum-mechanical transition in the classical theory, where the potential barriers are impenetrable, the transition {2}→{3} is just the intersection of the potentials C'.

The reported features of the transitions in quantum and classical treatments are fairly common, although they were first considered in some special cases [103,104]. We write the explicit form of the transition matrix element in the form [104]

$$\tilde{J}_{23} = 2(2\pi\Omega)w, \tag{5.36}$$

where $w = \dfrac{1}{4}\left(B_p - 2B_{p-1} + B_{p-2}\right)$ is the permeability barrier

$$B_p = (2k_\alpha)^p \, 2\exp(-2k_\alpha), \quad k_\alpha = \frac{\Omega}{4}m\Delta_x^2,$$

$p = 1/2$, $\Delta_x = |x'-x''|$ is the shift of the equilibrium position of a particle of the subsystem {3} with respect to the equilibrium position of the CPs. An additional factor of 2 was introduced in (5.36) compared with [43, 44]; this is associated with two versions of the transition: in the complete systems of the potentials well U_3 can be positioned both on the right of U_2 and to the left of it. Because in what follows we restrict the study to the movement of light atomic particles (protons, α-particles) with energy $E_0 \sim 1$ MeV, then according to estimates $k_\alpha \gg 1$. Therefore, at w it is sufficient to retain only the first term.

So, given the explicit form of the transition matrix element (5.36), the rate coefficient (5.28) in the case of weak interaction $\gamma_s^2 \ll \Omega^2$ has the form

$$R = \left(2\omega_{s_0}\right)^{-1}\left|\tilde{J}_{23}\right|^2 v\left(F_{21}\right)\sum_{s',s''} B_{s'}\gamma_{s''1}F_1\left(-s'',1;r'\right), \tag{5.37}$$

where

$$v\left(F_{21}\right) = \left\{\exp\left(F_{21}\omega_{s_0}\right)+1\right\}^{-1}.$$

Next, we calculate the relaxation frequency $1/\tau_e$ (5.27). In the case of light atomic particles with energies of 1 MeV velocity v_0 is always greater than Fermi velocity v_F. Therefore, when calculating the sum in (5.27) we restrict ourselves to the case $v_0 > v_F$ and, moreover, we neglect the thermal 'smearing' of the distribution n_k, considering the electron gas in the ground state.

Three types of diagrams in the k-th space, describing the various versions of particle scattering on free electrons, taking into account the Pauli exclusion principle, as required by condition $P_{k_1 k_2} \sim n_{k_1}(1-n_{k_2})$ were considered in [92]. In the calculation of $1/\tau_e$ we use a diagram [92] for the case $v_0 > v_F$, including the integration limits and the area allowed for angles $\mathbf{k}_1$ and $\mathbf{k}_1 - \mathbf{k}_2$. If we assume that all the momenta Δ_y and energy $Q(\Delta_y)$ are transferred to one electron, then $Q(\Delta_y) = \Delta_y^2/2m_e$. Then in the calculation of $1/\tau_e$ it is more convenient to relate $1/\tau_e$ the scattering cross section $d\sigma(\Delta_y)$ to the element $dQ(\Delta_y)$ and not to the element of the solid angle. Then, in high-velocity conditions we have

$$\frac{1}{\tau_e} = \Omega^2 \left(\frac{1}{v_F}\right)^2 \frac{1}{\tau_2(v_0)},$$

(5.38)

$$\frac{1}{\tau_2(v_0)} = \frac{2a_{TF}^2}{\Omega} \int d\sigma(\Delta_y)\pi(\Delta_y,\Omega)Q(\Delta_y).$$

The polarization operator in (5.38) has the form

$$\pi(\Delta_y,\Omega) =$$

$$= \frac{q_{TF}^2}{4\pi}\left\{\frac{1}{2} + \frac{k_F}{4\Delta_y}\left[\left(1 - \frac{\left(\Omega - \Delta_y^2/2m_e\right)^2}{\Delta_y^2 v_F^2}\right) \times\right.\right.$$

$$\times \ln\left|\frac{\Omega - \Delta_y v_F - \Delta_y^2/2m_e}{\Omega + \Delta_y v_F - \Delta_y^2/2m_e}\right| +$$

(5.39)

$$\left.\left. + \left(1 - \frac{\left(\Omega + \Delta_y^2/2m_e\right)^2}{\Delta_y^2 v_F^2}\right)\ln\left|\frac{\Omega + \Delta_y v_F + \Delta_y^2/2m_e}{\Omega - \Delta_y v_F + \Delta_y^2/2m_e}\right|\right]\right\}.$$

Due to the complex form of (5.39), the analytical expression for $1/\tau_2(v_0)$ (5.38), which is obtained by integrating over $dQ(\Delta_y)$, becomes cumbersome. In this regard, the polarization operator can be replaced by a quasi-static limit of $\pi(\Delta_y, 0)$ (Fig. 5.2). However, considerable simplification can be achieved in a more crude approximation. Indeed, comparison of the analytical results with the data obtained

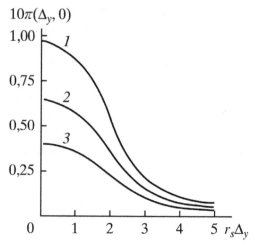

Fig. 5.2. The polarization operator (in Lindhard's form) for three values of parameter r_s: 2 (curve 1), 3 (curve 2) and 5 (curve 3).

in numerical integration shows that the polarization operator in the integrand of the form (5.38) can be approximated by its value at some point in the integration interval $(\omega_p/v_F) \leq \Delta y \leq 2m_e v_0$, namely the value $\pi(0,0)/1.6$.

Using this approximation, we find that

$$\left(\frac{1}{\tau_2(v_0)}\right) = \chi_2 \Psi_2(v_0), \quad \chi_2 = \frac{5}{8}\frac{1}{\Omega}\left(Z_1 e^2\right)^2 Z_V n_e d,$$

$$\Psi_2(v_0) = \frac{1}{10} L_e \theta_e(v_0),$$

(5.40)

where $\theta_e(v_0)$ is the minimum angle of electron scattering [106]. An expression was determined for $1/\tau_e$ (5.38) in the case of Rutherford scattering, when a collision is considered without taking into account the effect of screening.

5.6.2. Dechanneling due to the resonance transitions

At the depth of penetration of the order of the coherence length $L_{coh} \sim 10^4$ Å the quasi-equilibrium distribution has the form $n(\varepsilon_\perp) = \lambda \exp(-F_{21}\varepsilon_\perp)$ [107,108], where λ is the the absolute activity. As for the full distribution of the CPs, in addition to $n(\varepsilon_\perp)$, it includes $G(\varepsilon_\perp)$ – the density of states with the transverse energy in the interval $(\varepsilon_\perp, \varepsilon_\perp + d\varepsilon_\perp)$, so that

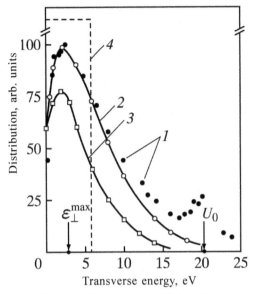

Fig. 5.3. The full distribution of the particles of the directional beam in respect of transverse energies.

$$p(\varepsilon_\perp)d\varepsilon_\perp = G(\varepsilon_\perp)n(\varepsilon_\perp)d\varepsilon_\perp. \qquad (5.41)$$

The solid points in Fig. 5.3 (number 1) are the results of numerical calculation of the density in a wide range $\varepsilon_\perp$ using equation (4.11). Curve 2 is the function $G(\varepsilon_\perp)$ for the CPs which, as shown, has a maximum at $\varepsilon_\perp = \varepsilon_\perp^{max}$ and therefore, $p(\varepsilon_\perp)$ (curve 3, where $1/F_{21} = 10$ eV) also has a peak shifted relative to the maximum $G(\varepsilon_\perp)$ to lower energies. Distribution $p(\varepsilon_\perp)d\varepsilon_\perp$ differs significantly from zero only in areas of low $\varepsilon_\perp$. With this in mind, we seek the solution of equation (5.26) for $r' = (\varepsilon_\perp/2\Omega) < 1$.

The results of the study of diffusion in configuration space [109] indicate that the shape of the initial distribution of particles has little effect on the diffusion fluxes. Much more important from this point of view is the depth at which the maximum of the initial distribution is found. Similarly, im the analysis of diffusion in the $\varepsilon_\perp$-space it is important to take into account the location of the maximum density of $G(\varepsilon_\perp)$. This can also be achieved in a simplified form of calculation by approximating the density of states by the dimensionless exponential function of the form

$$G(\gamma) = \begin{cases} G(0) \text{ for } \gamma \leq \gamma_1, \\ 0 \quad \text{ for } \gamma > \gamma_1, \end{cases} \tag{5.42}$$

where $\gamma = \varepsilon_\perp / U_0$, $\gamma_1 = \varepsilon_\perp' / U_0$ (function $G(\gamma)$ is shown by line 4 in Fig. 5.3). If we choose $\varepsilon_\perp' = 2\varepsilon_\perp^{max}$, then the normalization condition of the CPs per unit should be $G(0) = 1/\gamma_1$.

The expression for $G(\gamma)$ (5.42) was obtained on the basis of the Liouville equation with no contribution of inelastic scattering (see Section 4.2.3). This approach considers only the fluid flow in the transverse energy space, so $G(\varepsilon)$ corresponds to mechanical S-contributions to equation (4.1).

The degenerate hypergeometrical function, included in the equation (5.28), can be written as a series of Bessel functions $J_n(x)$. In order to ensure fast convergence of the series in s'', we use the same method as in Section 5.5.2 by replacing s'' by the root of the transcendental equation μ_r, whose form is defined below.

In the area of the transverse energy $\varepsilon_\perp < 2\Omega$ the value $(1/2 + \mu_r) \times (\varepsilon_\perp/2\Omega)$ is limited and in the expansion $_1F_1(-\mu_r, 1, \varepsilon_\perp/2\Omega)$ in (5.28) we restrict ourselves [110] to the 'older' term. Then, the equation can be written in a simpler form

$$2n(\varepsilon_\perp) \exp\left\{ -\frac{\varepsilon_\perp}{4\Omega} \right\} = \sum_r B(\mu_r) J_0\left(2\sqrt{\frac{\varepsilon_\perp \mu_r}{2\Omega}} \right). \tag{5.43}$$

Equation (5.43) contains only the zero-order Bessel function. Given that the model adopted in the distribution $n(\varepsilon_\perp)$ in the semi-classical approximation should be considered as 'truncated' when the value of the threshold energy $\varepsilon_\perp^* = U_0 + \varepsilon'$ (here $\varepsilon' > 0$), as well as taking into account (5.43), the equation for roots can be written as

$$J_0\left(2\sqrt{\frac{\varepsilon_\perp \mu_r}{2\Omega}} \right)\Bigg|_{\varepsilon_\perp = \varepsilon_\perp^*} = 0.$$

The solution of this equation is obvious:

$$\mu_1(\zeta) = \frac{1}{2} j_{01}^2 \zeta, \quad \mu_r(\zeta) = \mu_1(\zeta) \left(\frac{j_{0r}}{j_{01}} \right)^2, \tag{5.44}$$

where $\zeta = (\Omega/\varepsilon_\perp^*) \ll 1$, j_{0r} is the r-th root of the Bessel function of zero order. Thus, the summation in (5.43) is carried out over an infinite set of real roots $\mu_r(\zeta)$ (5.44), and $B(\mu_r)$ are the coefficients in the expansion of a given function in a Fourier–Bessel series. Using the general expression of the Fourier–Bessel theory, we get

$$B(\mu_r) = 4\lambda \left[J_1(j_{0r}) \right]^{-2} \mu_1 \left(\frac{2}{j_{01}} \right)^2 (1 + 4\Omega F_{21})^{-1} \exp(-2\mu_r). \quad (5.45)$$

Let us proceed directly to the calculation of R_e. As in the calculation of B_s, we replace the index s'' in (5.37) by the discrete value $\mu_r(\zeta)$. Passing to dimensionless variables, we obtain

$$R_e = \left| \tilde{J}_{23} \right|^2 \frac{1}{(2\Omega)^2} v(F_{21}) \frac{1}{\tau_e} \zeta^3 \times$$

$$\times \sum_r B(\mu_r) \mu_r \int_0^d d\eta\, G(\eta)\, {}_1F_1\left(-\mu_r, 1; \frac{\eta}{2\zeta} \right). \quad (5.46)$$

Substituting $B(\mu_r)$ (5.45) to R_e (5.46) and using (5.42), we integrate the resulting expression with respect to $d\eta$. We then introduce the the relaxation frequency $1/\tau_e$ (5.38) and (5.40) into the rate constant. The result is

$$R_e = 2^{5/2} \left| \tilde{J}_{23} \right|^2 k_3(F_{21}) \frac{1}{\tau_2(v_0)} \zeta^4 \left(\frac{1}{v_F} \right)^2 \left(\frac{1}{\gamma_1} \right)^{3/4}. \quad (5.47)$$

The value that is included in R_e does not depend on the particle density

$$k_3(F_{21}) = v(F_{21}) \sum_r \eta_r(\zeta) \quad (5.48)$$

and characterizes the kinetics of dechanneling. Function $\eta_r(\zeta) = \zeta j_{0r}^{3/2} \exp(-\zeta j_{0r}^2)$ contains an exponential factor decreasing with increasing r. This factor provides faster convergence of the series in (5.48), indicating the efficiency of the approach that we used in deriving (5.43) and (5.46) to improve the convergence of the series.

In studies of the channeling effect by the methods of quantum theory, we in fact used a 'hybrid' approach in the sense that the fast particles are considered as a classical subsystem, and the thermostat – as a quantum. A typical example is the study [79]. In contrast to this study, in the theory described previously the transition {2}→{3} does take place over the potential barrier and it occurs by the quantum-mechanical resonance in the excited state s_0. Reaching the level ω_{s_0} by the particle is the result of random walks between the levels of ω_s which are also considered in the framework of quantum theory.

This formulation of the elementary act of dechanneling is reflected in the structure of the rate coefficient, which includes the following qualitatively new elements:

– At $k_\alpha \gg 1$ with the accuracy to the constant factor (of the order of unity) the matrix element of the transition $\{2\} \rightarrow \{3\}$ is equal to the tunneling matrix element [111] between the states of the two-well zone, consisting of two harmonic wells.

– The transition matrix element (5.36) enters the dechanneling rate coefficient (5.47) explicitly, while a rough approximation is used in Section 5.5.2; the transmission coefficient of the barrier is replaced by a constant.

– The calculation took into account the scattering of fast particles in the one-particle excitations of the electron gas in the ground state. The relaxation constant (5.38) is expressed using a standard scattering cross section in the case of high particle velocities and the polarization operator. This formulation of the relaxation frequency is directly connected with the relations of the van Hove theory of irreversible processes [94].

We note one important fact. Considering the particles of both fractions of the directional beam as 'hot' atoms, we obtain the result (5.37), in which the kinetic coefficient contains an additional factor that depends on the transverse quasi-temperature $v\,(F_{21})$. This factor has no counterpart in any channeling theory, based on the principles of mechanics, nor in the classical Arrhenius kinetics. The function $v(F_{21})$ is a consequence of using the non-equilibrium statistical operator.

Directional beam separation leads to the separation of quasi-channeled particles into a special fraction. The particles of this fraction are in the ground state, moving into the area with high density of electrons and, in addition, they are heavily exposed to the thermal vibrations of atoms of the wall. Naturally, the energy losses of quasi-channeled particles, as noted above, are much higher than those of well-channeled particles. The difference between the losses of the particles of the subsystem $\{3\}$ and $\{2\}$ could be observed experimentally only in one case [112].

6

Non-equilibrium statistical thermodynamics of directional particle beams Concept of transverse quasi-temperature

6.1. Formulation of the theory based on expansion of the non-equilibrium statistical operator

The concept the quasi-temperature of fast particles, much higher than the thermostat temperature, was used to analyze a number of physical phenomena: cooling of a beam of fast protons in an electron gas [113], motion of cascade particles in the crystals in the dynamic phase [114] and the thermalization stage [115,116], thermalization of the relativistic positrons in solids [117,118], motion of fast electrons in a semiconductor in an external electric field [119]. Less commonly, this term was used in the theory of the channeling effect [16]. This section addresses the question of 'warming up' of high-energy atomic particles (protons, α-particles, etc.), moving in a crystal in the channeling mode.

Recall that at angles less than critical, the angular distribution of fast particles is divided into two parts: directed (aligned), and chaotic. In the quantum particle theory of particles, two parts of the beam correspond to sub-barrier (inside the channel) and above-barrier (outside the channel) transverse energy levels. Separation of the angular distribution and the formation of two groups of states occurs at a depth of penetration L of the order of coherence length L_{coh}, and most of the analytical calculations are carried out using the idea of establishing a quasi-equilibrium at this depth in the space of transverse momenta of the particles.

Indeed, at $L \sim L_{coh}$ the equation of motion for the density matrix of the CPs retains only the diagonal matrix elements that characterize the population of individual levels of the transverse energy of the particles. Therefore, at a depth $L \sim L_{coh}$ the equilibrium distributions for both the CPs and the chaotic particles are obtained. We restrict ourselves to the calculation of the basic thermodynamic parameter of the CP subsystem in the quasi-equilibrium conditions, namely, the steady-state values of quasi-temperature; the process of formation of quasi-temperature at small depths $L \lesssim L_{coh}$ as well as its subsequent evolution at the stage of thermalization of the beam is not considered.

As in the previous sections, the interaction between the CPs is not taken into account and the full Hamiltonian of the system can be written as a sum of two terms $H = H_1 + H_2$, namely, the Hamiltonian of free electrons $H_1 = H_e$ and CP Hamiltonian $H_2 = H_0^{(2)} + H_{int}^{(2)} + \tilde{H}_{int}^{(2)}$. Here $H_0^{(2)}$ includes the effective potential of the channel, describing the impact of a regular lattice on the CPs in the correlated collisions, $H_{int}^{(2)} = W_c^{(2)}$ corresponds to the close interaction of the CPs with the electrons, $\tilde{H}_{int}^{(2)}$ is the CP interaction with the plasmons. Using these model representations, we attempt to analyse the 'warm up' of the CP subsystem through quasi-elastic electron scattering.

The entropy operator and the entropy production operator are

$$S(t,0) = \Phi_1 + \sum_{mi} P_{mi} F_{im}(t),$$

$$\dot{S}(t,0) = \sum_{mi} \left\{ \dot{P}_{mi} F_{im}(t) + \left(P_{mi} - \langle P_{mi} \rangle_l^t \right) \dot{F}_{im}(t) \right\}. \tag{6.1}$$

Here Φ_1 is the Massieu–Planck functional,

$$\langle P_{mi} \rangle_l^t = \mathrm{Sp}(\rho_l P_{mi}), \quad \rho_l = \exp\{-S(t,0)\}.$$

We exclude from $\dot{S}(t,0)$ (6.1) time derivatives $F_{im}(t)$ using the equation

$$\sum_n \frac{\delta^2 S}{\delta \langle P_{mi} \rangle_l^t \delta \langle P_{ni} \rangle_l^t} \langle \dot{P}_{ni} \rangle_l^t = \dot{F}_{im}(t), \tag{6.2}$$

where $S = \langle S(t,0)\rangle^t$ is entropy. Then, by introducing energy and momentum fluxes of the CPs, taking into account the definitions (6.1), (6.2), after some transformations we obtain

$$\dot{S}(t,0) = \Delta\left\{ (F_{21} - \beta_1)\dot{H}_2 - \right.$$

$$-\sum_\alpha F_{21} V^\alpha \dot{\mathcal{P}}_2^\alpha - \frac{F_{21}}{mN_2}\sum_\alpha \left[\mathcal{P}_2^\alpha - mN_2 V^\alpha\right]\left\langle \dot{\mathcal{P}}_2^\alpha\right\rangle_I^t \right\} -$$

$$-\Delta\left\{\left[\tilde{H}_2 - \mu_2 N_2 - \frac{N_2\left(N_2, \tilde{H}_2 - \mu_2 N_2\right)^t}{\left(N_2, N_2\right)^t}\right] \times\right.$$

$$\times\left\langle \dot{H}_2 - \sum_\alpha V^\alpha \dot{\mathcal{P}}_2^\alpha\right\rangle_I^t \frac{F_{12}^2}{C_2}\right\}. \tag{6.3}$$

In (6.3) we use the notation: $\Delta A = A - \langle A\rangle_I^t$, μ_2 is the quasi-chemical potential of the CPs,

$$\tilde{H}_2 = H_2 - \sum_\alpha V^\alpha \mathcal{P}^\alpha + \frac{mV^2}{2}N_2, \quad C_2 = F_{21}^2\left(\tilde{H}_2 - \mu_2 N_2; \tilde{H}_2 - \mu_2 N_2\right)^t$$

is the specific heat of the CPs, $1/\beta_1$ is the temperature of the thermostat

$$\left(P_{mi}, P_{nj}\right)^t = \int_0^1 d\tau\, \mathrm{Sp}\left\{P_{mi} e^{-\tau S_0(t,0)}\left[P_{nj} - \langle P_{nj}\rangle_I^t\right]e^{(\tau-1)S_0(t,0)}\right\} \tag{6.4}$$

is the correlation function, $\dot{P}_{mi} = (1/i)\left[P_{mi}, H\right]$.

Assuming the interaction of electrons with the CPs is weak, we expand the non-equilibrium statistical operator

$$\rho(t) = \exp\left\{-S(t,0) + \int_{-\infty}^0 dt'\, e^{\epsilon t'}\dot{S}(t+t',t')\right\} \quad (\epsilon \to +0),$$

with respect to the powers of the interaction potential, with the form $\dot{S}(t,0)$ (6.3) taken into account. Keeping in this expansion only the first-order terms, the statistical operator can be written as

$$\rho(t) = \rho_I(t) +$$

$$+ \int_{-\infty}^{0} dt' e^{\varepsilon t'} \int_{0}^{t} d\tau e^{-\tau S_0(t,0)} \Delta \left\{ \left[F_{21} (t+t') - \beta_1 \right] \dot{H}_2(t') - \right.$$

$$\left. - \sum_{\alpha} F_{21}(t+t') V^{\alpha}(t+t') \dot{P}_2^{\alpha}(t') \right\} e^{(\tau-1)S_0(t,0)} -$$ \hfill (6.5)

$$- \int_{0}^{t} d\tau e^{-\tau S_0(t,0)} \Delta \left[\beta_1 \left(W_c^{(2)} + W_p^{(2)} \right) \right] e^{(\tau-1)S_0(t,0)},$$

where

$$S_0(t,0) = \Phi_0 + F_{21} \left(H_2 - \sum_{\alpha} V^{\alpha} P_2^{\alpha} - \tilde{\mu}_2 N_2 \right) + \tilde{B},$$

$$\Phi_0 = \ln \mathrm{Sp} \, \exp \left\{ -F_{21} \left(H_2 - \sum_{\alpha} V^{\alpha} P_2^{\alpha} - \tilde{\mu}_2 N_2 \right) - \tilde{B} \right\},$$

where $\tilde{\mu}_2 = \mu_2 - \frac{1}{2} m V^2$; $\tilde{B} = \beta_1 (H_1 - \mu_1 N_1)$, μ_1 is the chemical potential of electrons.

6.2. The energy–momentum balance equation in the comoving coordinate system

We assume that initially the fastest particles move along the crystallographic planes or axes in the z-direction with the mass velocity v_0. In order to eliminate directional movement, we transfer to the coordinate system moving with the particles. The equations of motion of the operators P_{mi}, involving only the transverse components of the physical quantities, can be written in the form of

$$\dot{H}_2 = \dot{H}_{2(c)} + n_2 \sum_{\alpha} \mathcal{F}^{\alpha} V^{\alpha}, \quad \dot{H}_1 = -\dot{H}_2,$$

$$\dot{P}_2^{\alpha} = \dot{P}_{2(c)}^{\alpha} + n_2 \mathcal{F}^{\alpha}, \quad \dot{P}_1 = 0, \quad \dot{N}_1 = 0,$$ \hfill (6.6)

where n_2 is the dimensionless density of the CPs; $\alpha = \{x, y\}$;

$$\dot{H}_{2(c)} = -i \left[H_0^{(2)}, W_c^{(2)} \right]; \quad \dot{P}_{2(c)} = -i \left[P_2, W_c^{(2)} \right].$$

The index c in (6.6) indicates the temporal changes due to close collisions; as regards the interaction with the plasmons, retardation of the CPs by interaction with a polarization cloud is described in terms of non-local forces

$$\mathcal{F}^\alpha = \frac{2e^2}{\pi^2} \int d\omega \int d\mathbf{q} \frac{1}{q^2} q^\alpha \operatorname{Im} \mathcal{E}^{-1}(\mathbf{q},\omega) \frac{1}{2}\delta\left(\omega - (\mathbf{q}\mathbf{V}) - \frac{q^2}{2M}\right),$$

where $\mathcal{E}(\mathbf{q},\omega)$ is the dielectric function of electrons.

Assuming that the energy exchange between the two subsystems is weak, we can restrict ourselves to the stationary version of the theory. In this case, to obtain the balance equations, it is sufficient to average the operator equations (6.6) over the distribution (6.5) in terms of stationarity. Thus, we find the balance equation for transverse energy

$$\sum_\alpha n_2 \mathcal{F}^\alpha V^\alpha + \int_{-\infty}^{0} dt'\, e^{\varepsilon t'} \left(\dot{H}_{2(c)}, \dot{H}_{2(c)}(t') \right)(F_{21} - \beta_1) -$$
$$- \sum_\alpha F_{21} V^\alpha \int_{-\infty}^{0} dt'\, e^{\varepsilon t'} \left(\dot{H}_{2(c)}, \dot{P}_{2(c)}^\alpha(t') \right) = 0 \tag{6.7}$$

and the transverse momentum balance equation

$$\mathcal{F}^\alpha n_2 + \int_{-\infty}^{0} dt'\, e^{\varepsilon t'} \left(\dot{P}_{2(c)}^\alpha, \dot{H}_{2(c)}(t') \right)(F_{21} - \beta_1) -$$
$$- \sum_\beta F_{21} V^\beta \int_{-\infty}^{0} dt'\, e^{\varepsilon t'} \left(\dot{P}_{2(c)}^\alpha, \dot{P}_{2(c)}^\beta(t') \right) = 0. \tag{6.8}$$

In accordance with the general principles of kinetic theory [121], in (6.7) and (6.8) we discard 'non-diagonal' correlation functions, since they contribute to higher order in the interaction potential in comparison with the contribution of 'diagonal' functions. For the same reason in terms of weak interaction, the 'diagonal' functions included in (6.7) and (6.8) will be calculated only in the Born approximation, although so far the balance equations have been written in more general terms. Within these approximations, based on the system of equations (6.7) and (6.8), we can easily obtain the balance equation of momentum energies. Indeed, excluding the non-local damping force in the equations (6.7) (6.8), we obtain

$$n_2 F_{21} \sum_{\alpha\beta} \left\langle V^\alpha V^\beta \right\rangle_0 D_{P_\alpha P_\beta} + (F_{21} - \beta_1) \left\langle (H_2)^2 \right\rangle_0 V_{\varepsilon\perp} = 0. \tag{6.9}$$

Here $\langle \ldots \rangle_0 = \operatorname{Sp} = (\rho_0 \ldots), \rho_0 = \exp\{-S_0(0,0)\}$,

$$D_{P_\alpha P_\beta} = \frac{1}{n_2} \int_{-\infty}^{0} dt'\, e^{\varepsilon t'} \left(\dot{P}_{2(c)}^\alpha, \dot{P}_{2(c)}^\beta(t') \right) \tag{6.10}$$

is the diffusion function of the CPs in the space of transverse momenta,

$$\nu_{\varepsilon_{\perp}} = Q \Big/ \left\langle \left(H_2 \right)^2 \right\rangle_0$$

is the relaxation frequency of transverse energy,

$$Q = \int_{-\infty}^{0} dt' \, e^{\varepsilon t'} \left(\dot{H}_{2(c)}, \dot{H}_{2(c)}(t') \right) \tag{6.11}$$

is the CP function. Also, keep in mind that equation (6.9) is obtained in the moving coordinate system, so the value $1/F_{21}$, incorporated in (6.9), should be interpreted as a thermodynamic parameter conjugate to the average transverse energy, i.e. as transverse quasi-temperature of the CPs.

6.3. Linearization of the balance equation
Contribution of energy dissipation (*V*-terms)

Linearization of the balance equation with respect to parameter F_{21} can be carried out using the following approximation. Averaging in the correlation functions, included in (6.10) and (6.11), can be performed on the quasi-equilibrium distribution of CPs in the space of transverse energies

$$p(\varepsilon_{\perp}) = G(\varepsilon_{\perp}) n(\varepsilon_{\perp}),$$

where $n(\varepsilon_{\perp}) \equiv n(\varepsilon_{\perp}; F_{21})$ is the average number of filled states with energies $\varepsilon_{\perp}$. The traditional theory of channeling (see, for example, [10,13,14,120]) was constructed in the approximation of constant values of $n(\varepsilon_{\perp}; F_{21})$ represented by its high-temperature limit. Interestingly, the same approximation of $n(\varepsilon_{\perp}; F_{21}) = $ const is used in the theory of motion of another class of fast particles – cascade particles [122]; a generalization of the cascade function in the case of finite parameter $1/F_{21}$ was considered in [123]. If in the calculation of correlation functions that are included in (6.10) and (6.11) we use the approach [120] and the averaging is performed using the non-Maxwellian function $G(\varepsilon_{\perp})$, then this would result in linearization of the balance equation for the transverse quasi-temperature of the CPs. (The density of states $G(\varepsilon_{\perp})$ is shown in Fig. 4.4.)

In addition, the balance equation can be considerably simplified if we move to the two-dimensional model of channeling in the crystal. Namely, we consider the motion of particles between two rows of atoms belonging to a square lattice in the plane (x, y). If we assume that the particle enters the lattice at an angle below the critical value, to some row of atoms oriented in the z-direction, its motion in the transverse direction (x-direction) is oscillating and in this case equation (6.9) retains only one diffusion function $\mathcal{D}_{p_x p_x} = \mathcal{D}(p_x)$. The latter can be calculated without direct calculation of the correlation function $\left(\dot{P}_{2(c)}^x, \dot{P}_{2(c)}^x (t') \right)$. Indeed, the probability $dw_1 (\mathbf{q}, \omega)$ that in unit time a particle passes to the electron gas momentum $\mathbf{q}$ and energy ω, is given by the standard formula of the theory of linear response

$$dw_1 (\mathbf{q}, \omega) = -\frac{1}{2} (4\pi)^3 |U(q)|^2 \, \mathrm{Im} \left\langle\!\!\left\langle \rho_\mathbf{q} | \overset{+}{\rho}_\mathbf{q} \right\rangle\!\!\right\rangle_\omega \times$$

$$\times \delta \left(\omega - (\mathbf{qV}) - \frac{q^2}{2m} \right) d\omega, \tag{6.12}$$

where $\left\langle\!\!\left\langle \rho_\mathbf{q} | \overset{+}{\rho}_\mathbf{q} \right\rangle\!\!\right\rangle_\omega$ is the Fourier component of the retarded Green function [44,58]; $U(q) = e^2 Z_1 / q^2$; $\rho_\mathbf{q}$ is the Fourier component of the operator of the electron density fluctuations. Knowing the probability (6.12), the diffusion function can be written as follows with the accuracy up to the second order of magnitude with respect to the interaction potential

$$D(p_x) = \frac{\tilde{\Omega}}{(2\pi)^3} \int d\mathbf{q} \int dw_1 (\mathbf{q}, \omega) q_x^2,$$

where, in accordance with conventional notions of the theory of scattering of particles on the electrons (see, for example, [124]) $\tilde{\Omega}$ is the unit volume in atomic units. Equation (6.12) shows that the effect of CPs on the electrons is similar to the effect of a test charge, whose frequency is equal to $(\mathbf{qV}) + q^2 / 2m$. Summarizing the diffusion function for the case of mechanical perturbations with multiple frequencies of excitation ω_{n0}, we obtain

$$\mathcal{D}(p_x) = \frac{\tilde{\Omega}}{(2\pi)^3} \sum_n \int d\mathbf{q} \, \tilde{w}_1 (\mathbf{q}, \omega_{n0}) q_x^2. \tag{6.13}$$

Here

$$\tilde{w}_1 (\mathbf{q}, \omega_{n0}) = -\frac{1}{2} (4\pi)^3 |U(q)|^2 \, \mathrm{Im} \left\langle\!\!\left\langle \rho_\mathbf{q} | \overset{+}{\rho}_\mathbf{q} \right\rangle\!\!\right\rangle_{\omega_{n0}}$$

is the corresponding probability, and $\left\langle\!\left\langle \rho_q \stackrel{+}{\rho_q} \right\rangle\!\right\rangle_{\omega_{n0}}$ describes the correlation of fluctuations of the electron density in the transition of the CP ground state (index 0) in the n-th excited state.

With regard to energy dissipation of the CPs, when calculating the function (6.11) it is important to take into account the fact that the scattering of electrons is not fixed at the center and it is fixed at the center of the scattering which oscillates between the channel walls. In other words, it is important to take into account the effect of electron dragging by a moving scattering center. First, we represent (6.11) the rate of change $\varepsilon_\perp$ as a result of transitions from the ground state of CP in all possible excited states

$$Q = \sum_n \left(\varepsilon_{\perp,n} - \varepsilon_{\perp,0} \right) w_{0n} = \mathrm{Re}\left\{ i \left\langle \left[H_1, W_c^{(2)} \right] \right\rangle_0 \right\}, \qquad (6.14)$$

where $\Delta\varepsilon_{n0} \equiv \varepsilon_{\perp,n} - \varepsilon_{\perp,0}$ is the energy transferred to electrons; w_{0n} is the probability per unit time, followed by energy transfer. Then, we move to a coordinate system fixed relative to the oscillating CP. Naturally, in this system, the scattering center is stationary and the interaction operator $\tilde{W}_c^{(2)}$ has the same form as that in the absence of movement of the CPs and, therefore, transforming the coordinates by a unitary transformation, we have

$$\exp\left(i\tilde{G}\right) W_c^{(2)} \exp\left(-i\tilde{G}\right) = \tilde{W}_c^{(2)}, \qquad (6.15)$$

where $\tilde{G} = -\left(\tilde{\Omega}\right)^{1/2} m_e \stackrel{*}{v}_{-q} \stackrel{*}{u}_x(t)$, m_e is the electron effective mass, v_{-q} is the Fourier component of the operator of the electron velocity, $u_x(t)$ is the lateral shift of the CP.

If we do not move in a moving system and the expansion of $W_c^{(2)}$ with respect to u_x we confine ourselves to the linear term, then, as the analysis shows, this leads to an additional condition imposed on the displacement $u_x \ll a_{\mathrm{TF}}$ (here, as usual, a_{TF} is the radius of the Thomas–Fermi screening). In reality in the channeling of fast particles, this condition is not satisfied and the transformation (6.15) allows us to move beyond the specified limit. Coordinate transformation in terms of dissipated energy for CP (6.14) is also achieved by a unitary transformation under the sign of the trace. The result is

$$\sum_n \left(\varepsilon_{\perp,n} - \varepsilon_{\perp,0}\right) w_{0n} = \mathrm{Re}\left\{\left\langle\left[\left[W_c^{(2)}, \tilde{G}\right], H_1\right]\right\rangle_{\tilde{G}} + \right.$$

$$\left. + i\left\langle\left[\left[\left[W_c^{(2)}, \tilde{G}\right], H_1\right], \tilde{G}\right]\right\rangle_0\right\}, \tag{6.16}$$

where $\langle...\rangle_{\tilde{G}} = Sp(\delta\rho_G ...)$, $\rho_{\tilde{G}} = \exp(-i\tilde{G})\rho_0\exp(i\tilde{G}) = \rho_0 + \delta\rho_G$. On the basis of the equation of motion for the statistical operator ρ_0 it is quite easy to obtain an exact equation satified by $\delta\rho_G$

$$\frac{\partial}{\partial t}\delta\rho_G = \frac{1}{i}[H, \delta\rho_G] + \frac{1}{i}[H_t, \rho_0] + \frac{1}{i}[H_t, \delta\rho_G], \tag{6.17}$$

where $H_t = i[H_1, \tilde{G}] + \partial\tilde{G}/\partial t$. Using (6.17), we calculate the non-equilibrium correction $\delta\rho_G$. Then, by averaging an arbitrary operator A with the distribution of $\rho_0 + \delta\rho_G$, we find the useful relation

$$\langle A(t)\rangle_{\tilde{G}} = \int_{-\infty}^0 dt' e^{\varepsilon t'} \langle\langle A(t) H_{t'}(t')\rangle\rangle. \tag{6.18}$$

If there is only one excitation frequency of the electron gas which is equal to the 'natural' frequency $\tilde{\omega}_{n0} = (\mathbf{qV}) + q^2/2m$, then the sum on the left hand side of (6.16) has only one term. In this case, which in fact is most interesting in terms of subsequent applications, the transition probability in terms of energy exchange between subsystems is calculated fairly simply.

Indeed, taking into account that the energy transfer oscillator is $\Delta\varepsilon_{n0} = (1/2)m\tilde{\omega}_{n0}^2 u_x^2$, and using (6.18), after some transformations we find that the transition probability $w_{0n} \equiv w_2(\mathbf{q}, \tilde{\omega}_{n0})$ has the form

$$w_2(\mathbf{q}, \tilde{\omega}_{n0}) = -2\pi\frac{1}{m\tilde{\omega}_{n0}}\mathrm{Im}\left\langle\left\langle f_\mathbf{q} \mid f_\mathbf{q}^+\right\rangle\right\rangle_{\tilde{\omega}_{n0}}, \tag{6.19}$$

$f_\mathbf{q} = -i\left[m_e v_\mathbf{q}, W_c^{(2)}\right]$ is the Fourier component of the fluctuation component of the force acting on the electrons from the CP. The drag of electrons in (6.19) can be expressed through the correlation of forces acting on the electrons. Summing the Fourier components of the probability and taking into account the energy conservation law, we obtain the desired frequency of the relaxation of the transverse energy

$$v_{\varepsilon_\perp} = -\tilde{\Omega}(2\pi)^{-2} \int d\omega \int d\mathbf{q} \frac{1}{m\omega} \mathrm{Im} \left\langle\left\langle f_q^+ | f_q \right\rangle\right\rangle_\omega \times$$

$$\times \delta\left(\omega - (\mathbf{qV}) - \frac{q^2}{2m}\right). \tag{6.20}$$

Thus, in analyzing the contribution of dissipative processes (V-terms) in the balance equation (6.9), we used a unitary transformation (6.15) in the same way as in (4.19). This allowed the dissipation of energy of the particles to be connected with the drag of electrons by scattering centers – channeled particles, oscillating between the walls of the planar channel.

6.4. The solution of the balance equation Transverse quasi-temperature of channeled particles (planar channel)

We determine the final form of the balance equation of the transverse momentum and transverse energy of the particles. Let's start with the planar channel. To do this, we generalize (6.9) for the case of the three-dimensional model of channeling, taking into account that $\mathcal{D}_{p_x p_x} = \mathcal{D}_{p_y p_y}$ [125]. Consider that even at low velocities of the CPs (for example, for protons with energies of 1 keV) the condition $(qv_0) \gg q^2/2m$ is fulfilled. However, we note that the averaging $\langle ... \rangle_0$ in the linearized equation (6.9) is reduced to averaging with the non-Maxwellian distribution function $G(\varepsilon_\perp) \, d\varepsilon_\perp$. The form of $G(\varepsilon_\perp)$ depends on the shape of the potential interaction of the CPs with the channel wall W_T. In particular, for a planar potential

$$W_T(x) = K\varkappa \, \mathrm{tg}^2 \frac{2\alpha x}{d_p} \simeq 4K\varkappa \left(\frac{\alpha x}{d_p}\right)^2 \tag{6.21}$$

density of states of CP can be written as [32]

$$G(\varepsilon_\perp) = \begin{cases} 0 & \text{for } \varepsilon_\perp < K\psi_0^2, \\ \dfrac{1}{2\pi\alpha} \dfrac{\varkappa^{1/2}(\varepsilon_\perp + \varkappa K)^{1/2}}{(\varepsilon_\perp + \varkappa K - K\psi_0^2)(\varepsilon_\perp - K\psi_0^2)^{1/2}} & \text{for } \varepsilon_\perp > K\psi_0^2, \end{cases} \tag{6.22}$$

where $K = \pi Z_1 Z_2 e^2 (Nd_p) d_p$, ψ_0 is angle of entry of the CP into the channel, d_p is the channel width, Z_2 is the effective number of electrons per lattice atom, Nd_p is the number of atoms per unit area of the plane of the channel. Given these observations and the equations (6.13) and (6.20), we write the balance equation in its final form:

$$2\pi n_2 \tilde{\Omega} \langle \varepsilon_\perp^k \rangle_0 4F_{21} \int_0^\infty d\omega \mid \int d\mathbf{q}\, q_x^2 |U(q)|^2 \times$$

$$\times \mathrm{Im} \left\langle\!\!\left\langle \rho_\mathbf{q} \mid \rho_\mathbf{q}^+ \right\rangle\!\!\right\rangle_\omega \delta\big((\mathbf{q}\mathbf{v}_0)-\omega\big)+$$

$$+(4\pi)^{-1}(F_{21}-\beta_1)\langle \varepsilon_\perp^2 \rangle_0 \int_0^\infty d\omega \int d\mathbf{q}\, \frac{1}{\omega} \times$$

$$\times \mathrm{Im} \left\langle\!\!\left\langle f_\mathbf{q} \mid f_\mathbf{q}^+ \right\rangle\!\!\right\rangle_\omega \delta\big((\mathbf{q}\mathbf{v}_0)-\omega\big)=0,$$

(6.23)

where $\varepsilon_\perp^k = E_0 \psi^2$ is transverse kinetic energy of the CPs, E_0 is the energy of longitudinal motion of the CPs, which is taken as the energy of entry into the channel, ψ^2 is the square of the angle between the direction of motion of the CPs and the plane of the channel.

Imagine also the explicit form of operators included in (6.23)

$$\rho_\mathbf{q} = \tilde{\Omega}^{-1/2} \sum_\mathbf{k} a_{\mathbf{k}-\mathbf{q}}^+ a_\mathbf{k},$$

$$f_\mathbf{q} = \tilde{\Omega}^{-1/2} \frac{i}{q} \sum_{\mathbf{k}\mathbf{q}'} (\mathbf{q}\mathbf{q}') U(q') \exp(i\mathbf{q}'\mathbf{R}) a_{\mathbf{k}-\mathbf{q}-\mathbf{q}'}^+ a_\mathbf{k},$$

(6.24)

and the explicit form of the imaginary part of the density–density Green function

$$\mathrm{Im} \left\langle\!\!\left\langle \rho_\mathbf{q} \mid \rho_\mathbf{q}^+ \right\rangle\!\!\right\rangle_\omega = \frac{1}{(2\pi)^2 q} 2\left(\frac{m_e^*}{}\right)^2 \int_0^\infty d\varepsilon \{n_e(\varepsilon+\omega)-n_e(\varepsilon)\}.$$

(6.25)

In (6.24), (6.25) we use the notation: $a_\mathbf{k}, a_\mathbf{k}^+$ are the Fermi operators; $n_e(\varepsilon)$ is the Fermi distribution for electrons; $\mathbf{R}$ is the radius vector of the CPs.

If the expression (6.24) is substituted in (6.23) and we perform integration over the angles of vector $\mathbf{q}$, the solution of the thus obtained equation has the form

$$\frac{1}{F_{21}} = \frac{1}{\beta_1}\left(1 + \frac{A}{B}\right),$$

$$A = 4\pi\left\langle\varepsilon_\perp^k\right\rangle_0 \int d\sigma(q)q\int_0^{qv_0} d\omega\left\{1 - \left(\frac{\omega}{qv_0}\right)^2\right\}\times$$

$$\times N(0)\frac{m_e}{k_{\mathrm{F}}}\int_0^\infty d\varepsilon\{n_e(\varepsilon+\omega) - n_e(\varepsilon)\}, \qquad (6.26)$$

$$B = \frac{1}{2(2\pi)^3}\left\langle\varepsilon_\perp^2\right\rangle_0 \int_0^\infty d\omega\frac{1}{\omega}N(0)\frac{m_e}{k_{\mathrm{F}}}\int_0^\infty d\varepsilon\{n_e(\varepsilon+\omega) - n_e(\varepsilon)\}\times$$

$$\times \int d\mathbf{q}' \int d\mathbf{q}''|U(q'')|^2\frac{1}{(q')^2}\frac{(\mathbf{q}'\cdot\mathbf{q}'')^2}{|\mathbf{q}'+\mathbf{q}''|}\delta\left(\mathbf{q}'\mathbf{v}_0 + \frac{(q')^2}{2m} - \omega\right).$$

Here

$$d\sigma(q) = \frac{8\pi Z_1^2 e^4}{v_0}NZ_2\frac{dq}{q^3}$$

is the scattering cross section; $N(0)$ is the density of electron states at the Fermi surface per unit volume of configuration space.

If the electrons are viewed as a degenerate gas at zero temperature, we can use the well-known [124] diagram of elastic ion–electron collisions in the momentum space. According to this diagram, the value of q' satisfies the following conditions:

$k_{\mathrm{F}} - v_0 m_e < q' < k_{\mathrm{F}} + v_0 m_e$ in the case of low velocities ($v_0 < v_{\mathrm{F}}$),

$m_e v_0 - k_{\mathrm{F}} < q' < m_e v_0 + k_{\mathrm{F}}$ in the case of high speeds ($v_0 > v_{\mathrm{F}}$).

Given these conditions and the Pauli exclusion principle, which actually amounts to the exclusion of the 'forbidden' areas on the collision diagram, we integrate with respect to $d\varepsilon$, $d\omega$ and the remaining angular variables in expressions A and B (6.26). As a result of cumbersome transformations we obtain the desired expression for the transverse quasi-temperature of the CP subsystem (planar channel)

$$\frac{1}{F_{21}} = \frac{1}{\beta_1}\left\{1 + \frac{5}{4}(4\pi)^3\frac{E_0\left\langle\psi^2\right\rangle_0}{\left\langle\varepsilon_\perp^2\right\rangle_0}\left(m_e v_{\mathrm{F}}\tilde\Omega\right)^{-1}\frac{I_1}{I_2}\right\}, \qquad (6.27)$$

where

$$I_1 = \int d\sigma(\theta)\sin^3\frac{\theta}{2}, \ I_2 = \int d\sigma(\theta)\left(\sin\frac{\theta}{2}\right)^{5/2} I_3(\theta),$$

$$I_3(\theta) = \int d\cos\theta\left(\cos^2\theta + \frac{4}{3}\right)\left(\cos\theta + \sin\frac{\theta}{2}\right)^{-1/2},$$

θ is the scattering angle, $\theta = \cos^{-1}\{(\mathbf{q}'\cdot\mathbf{q}'')/q'\cdot q''\}$.

So, in the moving coordinate system we calculate the main thermodynamic parameter of the CP – quasi-temperature (6.27). As can be seen from equations (6.8) and (6.10), the transverse momentum of the particles relaxes as a result of close collisions of the CPs with the conduction electrons and the consequence of this process is the increase in the transverse momentum of particles, accompanied by their slight braking.

This increase is equivalent to the diffusion of particles in the space of transverse momentum, and if we move from the space of transverse momentum into the space of transverse energies, we can see that as a result of the diffusion process the particles occupy higher and higher levels with increasing penetration depth. However, this process in the system is accompanied by a reverse process, namely, transverse energy dissipation, which is characterized by the function (6.11).

If the relaxation frequency of the transverse momentum is much greater than relaxation frequency $\varepsilon_\perp$, then because of the slow transfer of energy of fast particles to the thermostat the CP subsystem may be rapidly 'warmed up' and in this case quasi-temperature $1/F_{21}$ will exceed the electron temperature $1/\beta_1$. Interestingly, the same ratio of relaxation frequencies of energy and momentum exists in the theory of hot carriers in semiconductors [119], where the relaxation of energy requires a large number of collisions, and momentum relaxation occurs after a collision.

One more remark. In general, the stationary quasi-temperature in the subsystem {2} is established as a result of three processes: 1) diffusion of the CPs in the space of transverse energies, which leads to an increase in internal energy of the subsystem {2}, 2) dissipative process due to dragging of electrons, and 3) rapid internal thermalization of the subsystem {2} (mixing). The energy–momentum balance equation only includes the contribution of two processes: diffusion (heating subsystem {2} – the first term) and dissipation (cooling – second term).

As for internal thermalization, equation (6.9) does not take it into account, since in the limit of low density of the CPs the collisions between them can not be considered. This formulation of the problem coincides with the traditional formulation of the problem of Brownian particles, when we study the behavior of particles interacting only with atoms of the medium, but not with each other. In fact, the task in the theory of motion of hot carriers in semiconductors is formulated in the same manner [119].

Note that in the adopted model for complete description of the internal thermalization it is enough to take into account the effect of anharmonicity of vibrations of the particles between the channel walls [51,52]. The anharmonicity leads to a narrowing of the energy level spacing of the transverse energy with increasing number of energy levels. This leads to the transition of the energy of oscillations of one frequency to oscillations of another frequency, which, in turn, facilitates the exchange of vibrational energy between the CPs and, consequently, the establishment of quasi-equilibrium within the subsystem {2}.

The reality of the situation in which the transverse quasi-temperature of the CPs exceeds the temperature of the thermostat can only be explained specific experimental conditions. Since the value of $1/F_{21}$ (6.27) depends on the explicit form of functions $G(\varepsilon_\perp)$ and the specific values of the potential parameters W_T, we will describe two version of its evaluation. The first version will use the form $G(\varepsilon_\perp)$, which gives expression (6.22). As for the potential, as W_T is represented by the Lindhard potential in approximation by the (6.21).

Averaging the reduced energy $\epsilon_\perp = \varepsilon_\perp / K$ and the square of reduced energy $\epsilon_\perp^2$ with statistical weight $G(\epsilon_\perp)$, we obtain in the limit of the small angle of entry ($\psi_0 \rightarrow 0$)

$$\left\langle \epsilon_\perp \right\rangle_0 = \epsilon_\perp^{cr} \frac{\varkappa^{1/2}}{\pi} \left\{ \sqrt{1+\zeta} + \zeta \ln \left| \frac{1+\sqrt{1+\zeta}}{\sqrt{\zeta}} \right| \right\},$$

$$\left\langle \epsilon_\perp^2 \right\rangle_0 = \frac{\varkappa^{1/2}}{4\pi} \epsilon_\perp^{cr} \sqrt{\epsilon_\perp^{cr} \left(\epsilon_\perp^{cr} + \varkappa \right)} + \frac{3\varkappa}{4} \left\langle \epsilon_\perp \right\rangle_0,$$

(6.28)

where $\zeta = \varkappa / \epsilon_\perp^{cr}$, $\epsilon_\perp^{cr} = \varepsilon_\perp^{cr} / K$, $\varepsilon_\perp^{cr}$ is the critical transverse energy in the case of a planar channel. On the basis of (6.27), (6.28) and the formula for the effective mass in the Hartree – Fock approxmation

$$\frac{1}{2m_e^*} = \frac{1}{2m_e} + \frac{2}{3\pi m_e k_F a_B}$$

in the case of helium ions with energy $E_0 = 1$ MeV, moving in a planar channel (11) of a silicon crystal, we obtain the values of quasi-temperature given in Table 6.1.

When calculating $1/F_{21}$ the following values were also used: $\varepsilon_\perp^{cr} = E_0 \psi_p^2$; $\psi_p = \psi_{1/2}/0.6$; $\psi_{1/2} = 0.22°$; K = 225 eV, $d_p = 11.3 a_{TF}$, $1/\beta_1 = 300$ K; $\tilde{\Omega} = a_B^3$.

In the second version of the assessment of $1/F_{21}$ the density of states in the axial channel is determined. Let the area of the basal cell of the channel, which lies within the constant transverse potential energy $\Phi_\perp$, be equal to $f(\Phi_\perp)$. Then, the density (or normalized fraction) of the states of the CPs with full transverse energy $\varepsilon_\perp$, lower than the critical value $\Phi_\perp^{cr}$ [126], is

$$G(\varepsilon_\perp) = \frac{d}{d\varepsilon_\perp} f(\varepsilon_\perp - \tau_a). \tag{6.29}$$

Here $\tau_a = \tau_0 + \tau_s$, $\tau_0 = E_0 \psi_0^2$ is the initial transverse kinetic energy of the CPs, τ_s is the contribution of scattering at coherence depth. The function $f(\Phi_\perp)$ is determined by the potential relief of the CPs, and in the calculation of the contours of constant potential energy we use the potential of interaction of the CPs with the channel wall in the form [127]. In this case, the function $f(\Phi_\perp)$ can be approximated as

$$f(\Phi_\perp) = \frac{f_h}{\Phi_\perp^{cr}} \Phi_\perp,$$

where the value of $f_h = f(\Phi_\perp^{cr})$ depends on the geometry of the critical path, i.e. the contour corresponding to the saddle point of the potential relief of the CPs. Using (6.29) and linear approximation of the function $f(\Phi_\perp)$ (Fig. 6.1a), it is easy to show that the mean values included in (6.27) are equal

Table 6.1

Potential	α [22]	$\varkappa$ [22]	n_l (cm^{-2}) [23]	$\langle\epsilon_\perp\rangle_0$	$\langle\epsilon_\perp^2\rangle_0$	$1/F_{21}$, eV
Lindhard potential	1.10	0.07	$4 \cdot 10^{22}$	$2.5 \cdot 10^{-2}$	$3.5 \cdot 10^{-2}$	3
Moliere potential	0.86	0.15	$7.7 \cdot 10^{22}$	$4.8 \cdot 10^{-2}$	$10.4 \cdot 10^{-2}$	2

$$\left\langle \varepsilon_\perp^2 \right\rangle_0 = \frac{1}{3} f_h \left(\Phi_\perp^{cr} \right)^2, \ E_0 \left\langle \psi^2 \right\rangle_0 = \frac{1}{4} f_h \Phi_\perp^{cr}.$$

Now, on the basis of the thus obtained expressions we compute quasi-temperature (6.27) in the case of channeling of iodine ions in a silver single crystal of silver in the [011] direction. We obtain the following value

$$\frac{1}{F_{21}} = 5 \text{ eV}.$$

The following values were used [126]: E_0 = 21.6 MeV, $\Phi_\perp^{cr}$ = 60 eV, f_h = 0.31, v_F = 1.39 · 10^8 cm/s.

So, despite the different beams of CPs, various approximations of the distribution function and various forms of interaction potential W_T, the resultant quasi-temperature values similar in magnitude and fit well within the temperature range, typical for low-temperature plasma (one to several electron-volts). In both cases 'warming up' of the CPs takes place, so that at the coherence depth, in spite of achieving internal equilibrium in the CP subsystem, the system as a whole remains highly non-equilibrium with a large difference of the thermodynamic parameters of the CPs and the thermostat. High values of quasi-temperature (6.27) are a consequence of quasielastic scattering of the CPs on the electrons, when $v_{p_\perp} \gg v_{\varepsilon_\perp}$ (here $v_{p_\perp}$ is the relaxation frequency of the transverse momenta of the particles).

We conclude with two observations. First, the quasi-temperature which we considered as well as the quasi-temperature of other non-equilibrium systems, are of limited use. Namely, the value $1/F_{21}$ (6.27) is primarily a parameter that characterizes the statistical population of the levels of the transverse energy of fast particles. Secondly, the expression for quasi-temperature was obtained for the coupled system consisting of a thermostat and fast particles, and it is only natural that the thermostat temperature $1/\beta_1$ was used as the main thermodynamic parameter in (6.27). The presence in the theory of parameter $1/\beta_1$ suggests that the thermodynamic transition conditions transition

$$N_1 \to \infty, \ \Omega_1 \to \infty, \ \frac{N_1}{\Omega_1} = \text{const}$$

have been fulfilled (here Ω_1 is the volume of the thermostat and, consequently, of the system). Since these conditions are met, then the conditions for transition to the thermodynamics of the coupled system are also fulfilled.They also hold in the case when the CP subsystem includes only one particle, so that the value (6.27) preserves the meaning of the thermodynamic parameters even in the limiting case.

6.5. The dependence of equilibrium distributions on transverse quasi-temperature (axial channel)

6.5.1. The angular distribution of channeled particles after passing through a thin crystal

When the fast particles move in the axial channel their velocity is directed at a small angle to the crystallographic axis. The transverse kinetic energy of particles at the inlet and outlet of the thin crystal is denoted $\tau_\perp^{in}$ and $\tau_\perp^{out}$. Let the area of the basal cell of the channel, which lies within the contour of transverse potential energy $\Phi_\perp$, be equal to $f(\Phi_\perp)$. The energy $\Phi_\perp$ is not higher than the critical value $\Phi_\perp^{cr}$, and the contour of the critical potential energy Φ_h passes through a saddle point of the potential relief of the channel. The fraction of the cell area in the transverse plane within the contour Φ_h is denoted by f_h.

Assume that the thickness of the crystal is $\sim L_{coh}$. Then, the normalized fraction of states of the CPs with full transverse energy at output $\varepsilon_\perp^{out}$ and transverse kinetic energy in the range $(\tau_\perp^{out}, \tau_\perp^{out} + d\tau_\perp^{out})$ is

$$\frac{1}{f\left(\varepsilon_\perp^{out}\right)} f'\left(\varepsilon_\perp^{out} - \tau_\perp^{out}\right) d\tau_\perp^{out}.$$

Here we use the notation

$$f'\left(\varepsilon_\perp^{out} - \tau_\perp^{out}\right) = \frac{d}{d\Phi_\perp} f\left(\Phi_\perp\right)\bigg|_{\Phi_\perp = \varepsilon_\perp^{out} - \tau_\perp^{out}}.$$

It is assumed that in passage through the crystal the transverse energy of the CPs increases to $\varepsilon_\perp$ due to electron scattering. Then, the fraction of the states of the particles with the initial energy $\tau_\perp^{in}$ which received additional energy $\varepsilon_\perp$ and left the crystal, having the energy in the range $(\tau_\perp^{out}, \tau_\perp^{out} + d\tau_\perp^{out})$ and $(\varepsilon_\perp^{out}, \varepsilon_\perp^{out} + d\varepsilon_\perp^{out})$, is equal to

$$G\left(\varepsilon_\perp^{out}, \tau_\perp^{out}, \tau_\perp^{in}, \varepsilon_\perp\right) d\tau_\perp^{out} d\varepsilon_\perp^{out} =$$

$$= \frac{f'\left(\varepsilon_\perp^{out} - \tau_\perp^{out}\right)}{f\left(\varepsilon_\perp^{out}\right)} f'\left(\varepsilon_\perp^{out} - \tau_\perp^{in} - \varepsilon_\perp\right) d\tau_\perp^{out} d\varepsilon_\perp^{out}.$$

When the CP subsystem reaches quasi-equilibrium at the depth $\sim L_{coh}$, the probability of additional energy $\varepsilon_\perp$ being in the range $(\varepsilon_\perp, \varepsilon_\perp + d\varepsilon)$ is [128]

$$\rho(\varepsilon_\perp) = G(\varepsilon_\perp)n(\varepsilon_\perp),$$
$$n(\varepsilon_\perp) = \lambda e^{-F_{21}\varepsilon_\perp},$$
$$G(\varepsilon_\perp) = f'(\varepsilon_\perp - \tau_\perp) \qquad (6.30)$$

is the density of the states of the CPs, $\lambda = \exp(F_{21}\mu_2)$ is the absolute activity of the CPs. Absolute activity at high quasi-temperatures of the CPs is a slowly varying function and, if the distribution (6.30) is normalized to unity, it can be replaced by the limiting value of $\lambda = 1$ for $1/F_{21} \to \infty$. Averaging the normalized fraction of states with respect to the distribution (6.30), we obtain

$$\left\langle G\left(\varepsilon_\perp^{out}, \tau_\perp^{out}, \tau_\perp^{in}; \frac{1}{F_{21}}\right)\right\rangle d\tau_\perp^{out} d\varepsilon_\perp^{out} = \frac{f'\left(\varepsilon_\perp^{out} - \tau_\perp^{out}\right)}{f\left(\varepsilon_\perp^{out}\right)} \times$$
$$\times \int_0^\infty d\varepsilon_\perp \rho(\varepsilon_\perp) f'\left(\varepsilon_\perp^{out} - \tau_\perp^{in} - \varepsilon_\perp\right) d\tau_\perp^{out} d\varepsilon_\perp^{out}. \qquad (6.31)$$

Now we introduce the cumulative distribution of the CPs. We assume that the 'shoot-through' experiment is performed at $\tau_\perp^{out} = \tau_\perp^{in}$. Figure 6.1 shows as an example $f(\Phi_\perp)$ and $f'(\Phi_\perp)$ for a specific case: the axial channel of a [011] silver crystal, the fast particles – iodine ions [126, 129]. As can be seen from the figure, the fraction of the basal area of the cell $f(\Phi_\perp)$ and its derivative $f'(\Phi_\perp)$ can be approximated by a direct and a constant value, respectively. We use this result for the general case, writing $f(\Phi_\perp)$ in the linear approximation

$$f(\Phi_\perp) = \frac{f_h}{\Phi_h}\Phi_\perp.$$

In this approximation, using (6.30), the cumulative distribution of particles is equal to

$$\frac{1}{\Phi_h}\mathcal{F}\left(\Phi_h, \tau_\perp^{out}; \frac{1}{F_{21}}\right) \equiv \int_0^{\Phi_h} d\varepsilon_\perp' \left\langle G\left(\varepsilon_\perp', \tau_\perp^{out}, \tau_\perp^{out}; \frac{1}{F_{21}}\right)\right\rangle =$$
$$= \frac{f_h}{\Phi_h}\left[\ln\frac{\Phi_h}{\tau_\perp^{out}} - e^{F_{21}\tau_\perp^{out}}\Gamma\left(0, F_{21}\tau_\perp^{out}\right)\right], \qquad (6.32)$$

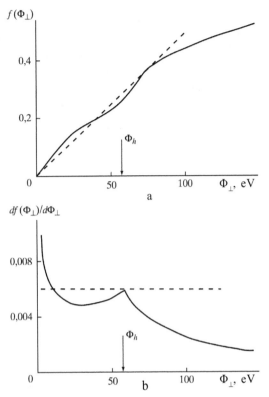

Fig. 6.1. Area of the cell of the axial channel inside the contour of transverse potential energy (a), the derivative of the cell area (b).

where $\Gamma(a, x)$ is the incomplete gamma function. Since $\tau_{\perp}^{out} = E_0 \theta_{out}^2$, formula (6.32) determines the distribution of the particles with respect to exit angles θ_{out} in geometry $\theta_{in} = \theta_{out}$.

Using (6.32), we calculate the peak height of the angular distribution of particles emerging from the back side of the crystal with the thickness $\sim L_{coh}$. To calculate the height, it is necessary to take into account the fact that the crystal plate is made up of individual blocks and has a mosaic structure with a finite angle of disordering of blocks ψ_m. Given the Gaussian distribution of the blocka, after averaging the expression (6.32) over the mosaic structure, we write the expression for the peak height as

$$I\left(\theta = 0; \frac{1}{F_{21}}\right) = \frac{\psi_{det}^2}{\psi_h^2} \int_0^\infty d\varphi \, \varphi \mathcal{F}\left(\Phi_h, E_0 \rho^2; \frac{1}{F_{21}}\right) \frac{1}{2\psi_m^2} e^{-\varphi^2/(2\psi_m^2)}. \quad (6.33)$$

Here $\Phi_h = E_0 \psi_h^2$, ψ_{det}^2 / ψ_h^2 is the geometrical factor, $\pi \psi_h^2$ is the limiting solid angle. We substitute (6.32), where $\tau_{\perp}^{out} = E_0 \varphi^2$, to the height of the peak (6.33) and integrate with respect to $d\varphi$. As a result, the height of the angular distribution can be represented as a function of the transverse quasi-temperature of the CPs

$$I\left(\theta = 0; \frac{1}{F_{21}}\right) = f_h \frac{\psi_{det}^2}{\psi_h^2} \left[0.28 + 0.5 \ln\left(\frac{\Phi_h}{2E_0 \psi_m^2}\right) + \right.$$

$$\left. + \left(2 - 4F_{21} E_0 \psi_m^2\right)^{-1} \ln\left(2F_{21} E_0 \psi_m^2\right)\right]. \tag{6.34}$$

The angular distribution of particles has a Gaussian shape [130, 131]. Therefore, we can write

$$I\left(\theta; \frac{1}{F_{21}}\right) = I\left(0; \frac{1}{F_{21}}\right) \exp\left\{-\frac{1}{1/F_{21}} E_0 \left[\theta - \sqrt{\left(\Delta\theta^2\right)_e}\right]^2\right\}. \tag{6.35}$$

Here

$$\left(\Delta\theta^2\right)_e = \frac{m_e}{2mE_0}\left(-\frac{dE}{dx}\right)_e L_{coh} \tag{6.36}$$

is the increase of the square of the angle of fluctuations due to electron collisions at the coherence depth.

Figure 6.2 shows the angular distribution of iodine ions after passing through a thin silver crystal calculated by formulas (6.34)–(6.36). (All parameters are given in Section 6.5.3.) As seen from Fig. 6.2, the form the theoretical curve agrees satisfactorily with the experimental data (points).

We make two observations. First, the dependence of the distribution (6.35) on quasi-temperature is determined by two factors: the exponential factor, which increases with $1/F_{21}$, and the height of the peak (6.34), which decreases with increasing $1/F_{21}$. Second, the Gaussian distribution corresponds to the case $\theta_{in} = \theta_{out}$.

But more complex geometry of the 'shoot-through experiment' can be used, namely, when $\theta_{in} \neq \theta_{out}$ and the output direction of the beam coincides with the direction of the atomic chain of one of the crystallographic planes of the crystal [37]. Then, the angular distribution has two peaks [37]: a large maximum in the direction $\theta_{in} = \theta_{out}$ and a small one – in the direction of the chain. The latter is associated with the effect of engaging the initially unchanneled particles in the regime of directed motion under the influence of the

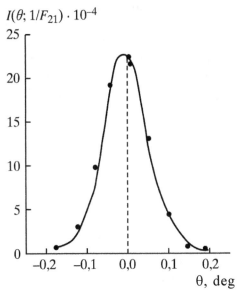

$I(\theta; 1/F_{21}) \cdot 10^{-4}$

Fig. 6.2. Angular distribution after passage of the crystal in the geometry of the 'shoot-through' experiment.

planar potential. The angular distribution of the particles involved in the regime is Gaussian.

Using a linear approximation for $f(\Phi)$, the general expression for transverse quasi-temperature (6.26) can be significantly simplified. We consider the electrons as a degenerate gas at zero temperature. Then, under the conditions of axial channeling for $v_0 > v_F$ after integration over $d\omega$, $d\varepsilon$, $d\mathbf{q}'$, we obtain

$$\frac{1}{F_{21}} = \frac{1}{\beta_1}\left\{1 + 20\pi^2 K \frac{\varepsilon_F}{k_F^3}\frac{I_1}{I_2}\right\}. \tag{6.37}$$

In the linear approximation, the value of K, included in (6.37), takes the simple form

$$K = \frac{\int_0^{\Phi_h} d\varepsilon_\perp \varepsilon_\perp \dfrac{d}{d\varepsilon_\perp} f\left(\varepsilon_\perp - \tau_\perp\right)}{\int_0^{\Phi_h} d\varepsilon_\perp \varepsilon_\perp^2 \dfrac{d}{d\varepsilon_\perp} f\left(\varepsilon_\perp - \tau_\perp\right)} = \frac{3}{2}\frac{1}{\Phi_h}. \tag{6.38}$$

6.5.2. Distribution of particles over exit angles in backscattering

We assume that the atomic chains, forming an axial channel, are 'frozen' and the rms scatter of the atoms in the transverse plane is equal to σ. We introduce following notation: θ – the angle between the velocity of the CP and the chain axis, ρ' – the distance from the particle chain axis in the transverse plane. Statistical mechanistic theory [132] established a link of the angular distribution $\tilde{I}(\theta)$ in the transverse kinetic energy $\varepsilon_\perp^k = E_0\theta^2$ with the distribution of total transverse energies $\tilde{P}\left(E_\perp(\theta,\rho')\right)$:

$$\tilde{I}(\theta) = 2\int_0^{\rho_0} d\rho'\rho' \frac{\tilde{P}\left(E_\perp(\theta,\rho')\right)}{\rho_0^2 - \Lambda^2(\rho')}. \tag{6.39}$$

Here, $E_\perp(\theta,\rho') = \varepsilon_\perp^k + U_a(\rho')$ is the transverse energy of the axial channel; $\rho_0 = (\pi N d)^{-1/2}$ is the radius of the area related to one atomic chain in the transverse plane, N is the density of the lattice of the atoms, d is the distance between the atoms in the chain. The value of $\Lambda(\rho')$, as shown in [132], is determined from the condition $U_a(\Lambda) = E_\perp$.

We calculate the density of states of the particles in the problem of backscattering. Separating the components of the radius vector and the velocity of the particles (ρ, z) and $(v_\perp, v_z)$, we write the probability density of finding particles in the interval $(\rho, \rho + d\rho)$ with velocities in the intervals $(v_z, v_z + dv_z)$ and $(v_\perp, v_\perp + dv_\perp)$. Let $\tilde{d}(\theta)$ be the depth from which the averaged chain potential U_a acts on the particle. Then, considering that at small depth (in the surface layer), the density of states has a Gaussian shape [32], it is easy to determine this density for a large depth by replacing ρ' by $\rho' + \left(v_\perp \tilde{d}(\theta)/v_z\right)$ in the equation for density [132]. Taking into account that in the backscattering problems parameter $\tilde{d}(\theta)$ can be replaced by its asymptotic value of $d/2$ [10], we have

$$G\left(\rho',v_\perp,v_z,\theta \mid z = \frac{d}{2}\right) = \left(2\pi\sigma^2 \cdot 2\pi v_0 \sin\theta\right)^{-1} \times$$

$$\times \exp\left\{-\frac{1}{2\sigma^2}\left(\rho' - \frac{v_\perp d}{2v_z}\right)^2\right\} \delta(v_\perp - v_0\sin\theta)\delta(v_z - v_0\cos\theta).$$

We now turn to the density of states in the variables θ, ρ'. For this, $G\left(\rho',v_\perp,v_z,\theta \mid \frac{d}{2}\right)$ must be integrated by $dv_\perp$, dv_z. We write the equation so obtained in the dimensionless form:

$$G\left(E_\perp(\theta,\rho')\right)=\sigma^{-2}\int_{\Lambda(\rho')}^{\rho_0} d\rho\, \rho \times$$

$$\times \exp\left\{-\frac{1}{2}\left[B^2(\theta,\rho,\rho')+(\rho/\sigma)^2\right]\right\}J_0(iz'),$$

(6.40)

where $J_0(x)$ is the Bessel function,

$$B^2(\theta,\rho,\rho')=\left(\frac{d}{2\sigma}\right)^2 \frac{1}{E_0}\{E_\perp(\theta,\rho')-U_a(\rho)\}, \quad z'=\frac{\rho}{\sigma}B(\theta,\rho,\rho').$$

With the explicit form of the density (6.40), it is easy to generalize the angular distribution on the basis of the definitions of statistical thermodynamics. For this purpose, we introduce the average occupation number of states with energy

$$n\left(E_\perp(\theta,\rho')\right)=\lambda e^{-F_{21}E_\perp(\theta,\rho')}$$

(6.41)

(this form is similar to (6.30)). Using (6.39), (6.40) and (6.41), we represent the desired angular distribution as a function of the transverse quasi-temperature:

$$\tilde{I}\left(\theta;\frac{1}{F_{21}}\right)=2\int_0^{\rho_0} d\rho'\, \rho' \frac{1}{\rho_0^2-\Lambda^2(\rho')}\times$$

$$\times G\left(E_\perp(\theta,\rho')\right)n\left(E_\perp(\theta,\rho')\right).$$

(6.42)

We transform (6.42) and pass to the dimensionless variables

$$\zeta=\frac{\rho}{a_0}, \quad \zeta_0=\frac{\rho_0}{a_0}, \quad \xi=\frac{\rho'}{a_0}, \quad \eta=\frac{a_0^2}{2\sigma^2},$$

$$\psi=\frac{\theta}{\psi_{cr}}, \quad \varkappa^2=\left(\frac{d^2}{8\sigma^2}\right)\left(\frac{U_0}{E_0}\right), \quad \gamma=\frac{R_{eff}}{a_0},$$

where a_0 is the screening radius; U_0 is the potential barrier height in the axial channel; ψ_{cr} is the critical channeling angle; $R_{eff}=\sqrt{a_0^2+2\sigma^2}$ is the effective radius of action with the displacement of the lattice atom from the site over distance σ taken into account. Next, we write down the chain potential as $U_a(\zeta)=U_0 w(\zeta)$ and in the first approximation we approximate $w(\zeta)$ by the step theta function $\Theta(\gamma-\zeta)$. After the transition to dimensionless variables, we substitute (6.40) into (6.42). As a result of transformations, the distribution of backscattered particles, which originally (in the channeling mode) reached a depth of $\sim L_{coh}$, takes the form

$$\tilde{I}\left(\psi;\frac{1}{F_{21}}\right)=2\lambda\eta\frac{1}{\zeta_0^2}e^{-F_{21}U_0\psi^2}\left\{\gamma^2e^{-F_{21}U_0}R_1\left(\psi\right)+2\tilde{R}_0\left(\psi\right)\right\}, \qquad (6.43)$$

where

$$R_n\left(\psi\right)=\int_0^{\zeta_0}d\zeta\,\zeta\Phi_n\left(\psi,\zeta\right),$$

$$\tilde{R}_n\left(\psi\right)=\int_0^{\zeta_0}d\xi\,\xi\left[\zeta_0-\zeta_1\left(\psi,\xi\right)\right]^{-1}\int_{\zeta_1(\psi,\xi)}^{\zeta_0}d\zeta\,\zeta\Phi_n\left(\psi,\zeta\right),$$

$$\Phi_n\left(\psi,\zeta\right)=\exp\left\{-\left[\eta\zeta^2+\varkappa^2\left(\psi^2-w(\zeta)+n\right)\right]\right\}\times$$

$$\times\sum_{m=0}^{\infty}\varkappa^{2m}\left(m!\right)^{-2}\left[\eta\zeta^2\left(\psi^2-w(\zeta)+n\right)\right]^m.$$

Function Φ_n (ψ, ζ) in (6.43) is presented as a series in powers of $\varkappa^2$, since at high energies $\varkappa^2\ll1$ the index takes on two values: $n=0$ and 1. The value of ζ_1 (ψ, ξ) in (6.43) is defined implicitly by

$$w\left(\zeta_1\left(\psi,\xi\right)\right)=\begin{cases}\psi^2+w(\xi) & \text{at }\psi^2+w(\xi)\le1,\\ 0 & \text{at }\psi^2+w(\xi)>1.\end{cases}$$

Only the first term is retained in the expansion of function $\Phi_n(\psi,\zeta)$. $R_n(\psi)$ and $R_n(\psi)$ are then integrated over $d\xi$, $d\zeta$ and these expressions are substituted into (6.43). The result is that the minimum yield of backscattered particles χ_{min} $(1/F_{21})=\tilde{I}\left(\psi=0;1/F_{21}\right)$ in actual physical conditions $\eta\zeta_0^2\gg1$ depends on the transverse quasi-temperature (6.37), (6.38) as follows:

$$\chi_{min}\left(\frac{1}{F_{21}}\right)=\lambda\left(\frac{\gamma}{\zeta_0}\right)^2e^{-F_{21}U_0}. \qquad (6.44)$$

As shown in [128], for the transition from statistical thermo-dynamics to the formulas of Lindhard's theory [10], it is sufficient to perform the limiting transition $1/F_{21}\to\infty$. At this limit, the formula (6.44) gives the expression

$$\chi_{min}=\left(\frac{\gamma}{\zeta_0}\right)^2=\frac{a_0^2+2\sigma^2}{\rho_0^2},$$

which completely coincides with the expression [10].

The distribution of the backscattered particles in a wide angle range can be calculated using equation (6.43) and by numerical integration. The results of these calculations are presented in Fig. 6.3 which shows the angular distribution of protons with the energy of

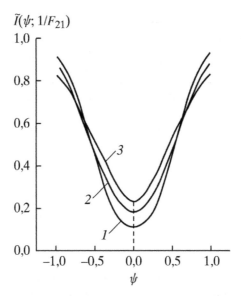

Fig. 6.3. Angular distribution of the backscattered particles at three values of the quasi-temperature of the beam.

1.6 MeV which initially travelled to the depth of $\sim L_{coh}$ in the axial channel of a [110] silicon crystal. The curves in the figure correspond to three values of quasi-temperature, namely: 1) $1/F_{21} = U_0/3$; 2) $1/F_{21} = 2U_0/3$; 3) $1/F_{21} = U_0$. According to Fig. 6.3, the minimum of the angular distribution decreases with decreasing $1/F_{21}$, which is in complete agreement with the analytical theory results (6.44).

6.5.3. Determination of transverse quasi-temperature based on experimental data

The transverse quasi-temperature of the CPs can be determined by experimental measurements, if we use the peak height of the angular distribution obtained in the 'shoot-through' experiment. Indeed, if the observed value of the peak height is substituted in expression (6.34), the latter is transformed into a transcendental equation for $1/F_{21}$ (all other parameters in a particular experiment are known). The solution of the transcendental equation gives the value $1/F_{21}$.

Figure 6.2 shows the angular distribution of the iodine ions passing through a silver crystal [126,129]: crystal thickness 0.835 mm, CP energy 12.6 MeV, axial channel [011], $(-dE/dx)_e =$ 3.7 MeV /μm, $f_h = 0.31$, $\psi_{det} = 0.012$ deg.

Table 6.2

Crystal thickness L, μm	Critical energy Φ_h, eV	Critical angle ψ_h, deg	Disordering angle ψ_m, deg	Peak height I (0; $1/F_{21}$)	Quasi-temperature $1/F_{21}$, eV
0.835	60	0.12	0.025	$21.9 \cdot 10^{-4}$	6

Using the experimental values of the parameters included in (6.34), we find that in the case of passage of iodine ions the transverse quasi-temperature of the beam is 6 eV. The remaining data are included in Table. 6.2.

6.6. Quasi-temperature of strongly non-equilibrium systems

6.6.1. Dependence of the coefficient of the dechannelling rate and dissociation of molecules in plasma on quasi-temperature

It is established that at the penetration depth of $\sim L_{coh}$, despite achieving internal equilibrium in the CP subsystem, the overall system is strongly non-equilibrium with a large difference in the thermodynamic parameters of the individual subsystems.

Primary processes of passage of the CPs through the crystal are acts of scattering on electrons and phonons. Each act is accompanied by a transition of the CPs between the levels of transverse energy. If in the process of random walks on the levels the CP reaches ω_{s_0}, corresponding to the capture potential of the particles in the channeling mode, the CP travels to the chaotic part of the beam, i.e. it is dechanneled. The act of dechanneling can be described using a 'quasi-chemical' reaction in a homogeneous, closed system

$$\begin{cases} S\{A_1\} + A_2 \underset{k_{-1}}{\overset{k_1}{\rightleftarrows}} S\{A_1\} + \overset{*}{A_2}, \\ \overset{*}{A_2} \overset{k_2}{\to} A_3, \end{cases} \tag{6.45}$$

where the symbol A_i corresponds to particles of i-th subsystem, and $S\{A_1\}$ to the set of atoms of the thermostat. According to reaction (6.45), in addition to the transition of the CPs to the level ω_{s_0} (rate constant k_1) dechanneling also includes the reverse process – relaxation with the rate constant k_{-1} and quasi-chemical $\{2\} \to \{3\}$ with constant k_2.

Since the density of atoms of the thermostat is many orders higher than the density of the CPs, it can be argued that a steady flow of reaction (6.45) starts fairly quickly [133]. Moreover, in the steady-state transitions {2}→{3} and relaxation are balanced by the formation of excited particles in the reaction with the rate constant k_1. Since the thermodynamic parameters of the thermostat settings are significantly different from those of the CPs (quasi-temperature estimate (6.37) in the case of channeling of iodine ions in the crystalgives $1/F_{21}$~5 eV), in terms of balancing mentioned above, the following relationship holds $k_1/k_{-1} = \exp(-F_{21}\omega_{s_0})$.

CP relaxation with rate constant k_{-1} takes place mainly by the resonant transition {2}→{3} (resonance energy), so that $k_2 \ll k_{-1}$. Given this condition, and the ratio of chemical kinetics for a two-step reaction [133], we can write the rate constant for reaction (6.45) as follows:

$$k_{tot} = k_2 \left(\frac{k_{-1}}{k_1} + 1 \right)^{-1}. \tag{6.46}$$

Substituting the explicit form of the relationship k_1/k_{-1} in (6.46), we obtain

$$k_{tot} = v(F_{21}) \equiv \left\{ e^{F_{21}\omega_{s_0}} + 1 \right\}^{-1}. \tag{6.47}$$

Thus, the rate constant (6.47) includes the transverse quasi-temperaturesu of the CPs as a function of $v(F_{21})$. The same function was included in the rate of dechanneling, derived from the generalized transport equations [87].

The range of applicability of the reaction (6.45) is not exhausted by the process of dechanneling. It can be used in the analysis of a wide range of processes due to the random walk of particles on the levels of vibrational energy between absorbing and reflecting barriers. In particular, the reaction (6.45) describes the dissociation of molecules in the plasma. The large deviation of the system from thermodynamic equilibrium in this case is due to incomplete vibrational relaxation which in turn leads to a significant separation of the temperature of the vibrational degrees of freedom from translational temperature [134].

Recall that in the theory of dissociation [134] the molecule is treated as an oscillator which as a result of collisions with other molecules (oscillators) undergoes transitions between the vibrational levels. If this molecule reaches a level with energy E_D equal to dissociation energy, the subsequent collision leads to a transition of its discrete spectrum into a continuous one, i.e. to the disintegration of the molecule.

Given the difference in the thermodynamic parameters of the theory of channeling and dissociation, two branches can be obtained from (6.47):

$$k_{tot} = \begin{cases} \dfrac{1}{2} \dfrac{1}{F_{21}\omega_{s_0}} \left(\dfrac{1}{2} + \dfrac{1}{F_{21}\omega_{s_0}} \right)^{-1} . \\ e^{-E_D/T_V} \end{cases}$$

(6.48)

The first branch corresponds to the dechanneling of protons ($E_0 \sim 1$ MeV, $2F_{21}\omega_{s_0} < 1$), second – the dissociation of molecules in the plasma ($F_{21}E_D \equiv (E_D / T_V) > 1$). Note that the second branch of the rate constants may further include a factor describing the lag of T_V in relation to translational temperature. In addition, under conditions of weak discreteness of the spectrum of transverse energy of the channeled particles the diffusion of the particles in the transverse energy space can be viewed as an ongoing activationless process. At the same time, the dissociation process (the second branch k_{tot}) is activated.

6.6.2. Quasi-temperature as quasi-equilibrium distribution modulus

Expressions (6.30) and (6.41) are the quasi-equilibrium distributions in a macroscopically small volume containing a fast particle at any time. In this volume, more precisely, within a block of coherent regions [79], the modulus of the established distribution is transverse quasi-temperature $1/F_{21}$ (6.27), (6.37). The rest of the volume of the crystal, excluding the block of coherent regions, remains in thermodynamic equilibrium with temperature $1/\beta_1$, and $1/\beta_1 \ll 1/F_{21}$. In this formulation of the problem there is no contradiction with the general principles of non-equilibrium thermodynamics [44]; the latter permits the presence im the system of two time scales of relaxation and, most importantly, the short time to establish quasi-equilibrium in a macroscopically small volume.

The system state can be described not only by a set of averages $\langle P_{mi} \rangle$ but also by a set of the variances of these quantities. The stationary value of quasi-temperature $1/F_{21}$ can in principle be easily related to the stationary variance of the transverse energy of the

CPs. However, we have chosen a different approach, based on the balance equation (6.9), making it possible to separately calculate the contribution of V-terms: diffusion of particles in the space of transverse energy ('heating' of the subsystem {2}) and the dissipative process (cooling subsystem {2}). If, however, we follow the path of direct calculation of the variance, the energy dissipation of the CPs due to entrainment of electrons by the scattering center oscillating in a channel [61] cannot be taken into account.

As for the dependence of the angular distribution on quasi-temperature, the maximum height (6.34) of the distribution of the particles passing through the crystal decreases with increasing $1/F_{21}$, and the minimum yield (6.44) in the backscattering conditions increases. Given the proportionality of the quasi-temperature of the CPs with the temperature of the thermostat, we can say that $I(0; 1/F_{21})$ decreases, and $\chi_{min}(1/F_{21})$ increases with increasing $1/\beta_1$. This temperature dependence was observed in experimental studies: in [135] – for the height of the peak, in [136,137] – for the minimum output.

Using the classification given in Section 4.1, the quasi-temperature of the CPs (6.27), (6.37) can be regarded as the thermostat temperature renormalized by taking into account the V-terms. As a result, the total distribution (6.30) includes the contribution of both V-terms included in $1/F_{21}$, as well as of S-terms that are included in $G(\varepsilon_\perp)$ (6.22), (6.29). The quasi-temperature of the fast particles is determined by the properties of the distribution of the entire ensemble as a whole; the quasi-temperature is not the mean value of any one-particle quantity. In particular, quasi-temperature is not the average of some microscopic dynamical feature, such as transverse energy.

The dependence of the functions appearing in the channeling theory on the quasi-temperature of the CPs has not been previously considered. The development of the concepts of non-equilibrium statistical thermodynamics allows us to conclude that in the quasi-equilibrium state these functions depend strongly on $1/F_{21}$. However, the main characteristic of the kinetic process (6.47) has a weaker dependence on the quasi-temperature of the CPs, especially when $F_{21}\omega_{s_0} < 1$.

6.7. Comparative analysis: channeled particles and the spin gas

6.7.1. Spin temperature in a strong magnetic field

Initially, the concept of the temperature of the subsystem of nuclear spins in the crystals was put forward as a hypothesis without any theoretical justification. In this case, the correctness of the claims depended on the successful interpretation of many experimental data. Therefore, the spin temperature concept was included in the final analysis in the general theory of magnetism of solids.

The simplest version of the introduction of the concept of spin temperature will be briefly discussed. Namely, the version in which it was introduced in the population of the Zeeman levels of the nuclear spins in strong magnetic fields will be discussed. We use the model of an ideal spin gas, according to which the spins interact with an external magnetic field, but do not interact with each other. In this model, when the spin subsystem $I^z = 1/2$ (here, as usual, I^z is z-component of the nuclear spin) reaches thermodynamic equilibrium with the lattice temperature $1/\beta_1$, the ratio of the populations of the Zeeman levels is

$$\frac{n_+^0}{n_-^0} = \exp\left(\beta_1 \gamma \tilde{H}\right),$$

where $\tilde{H}$ is the intensity of the strong magnetic field. The equilibrium with the lattice can be shifted in two ways: either by placing the spin subsystem in the RF field, or by changing the external magnetic field 'in a jump'. Both these options will be discussed below. After the violation of statistical equilibrium the spin subsystem can again be described using the ratio of the populations n_+ and n_-, if we assume that this ratio depends on temperature $1/F_s$ which is different from the lattice temperature. In the original versions of the theory this temperature is called the spin temperature. So, we have

$$\frac{n_+}{n_-} = \exp\left(F_s \gamma \tilde{H}\right). \tag{6.49}$$

Introducing the definition (6.49), we actually assume that in the energy representation the density matrix of the spin subsystem is diagonal. Since the off-diagonal matrix elements decay over the time of the order of the spin–spin relaxation time T_2, the spin temperature $1/F_s$ should be considered only when the condition

$T_2 \ll T_1$ is satisfied, where T_1 is the characteristic time of spin-lattice relaxation. The said condition is usually satisfied, since the spin-spin (dipole–dipole) interaction is stronger due to the lattice spins. Thus, at time intervals $T_1 > \Delta t > T_2$, in fact, we can use (6.49) and the concept of spin temperature.

Here it is appropriate to specify one more point. Interaction between nuclear spins in general can not be regarded as weak, so the components of the individual spins I^z can not be regarded as 'good' quantum numbers and a more consistent approach makes it necessary to analyze the solid as a single spin system. However, in strong magnetic fields, much stronger than the local field, i.e. Weiss field, it is nevertheless possible to use the model of individual spins with fairly well-defined energy levels [137]. Also, the concept of spin temperature is almost never used for low-temperature systems. Therefore, we restrict ourselves to studying only high-temperature cases. Dipole – dipole interaction between spins can be the starting point for several types of transitions, including the 'tilting' of two spins with opposite directions. In particular, in a rigid lattice 'tilting' of spins has a finite probability. Of course, the total Zeeman energy of spins is preserved in these transitions. We also note that in strong magnetic fields, the average energies of the Zeeman and spin-spin subsystems are of comparable magnitude, so the study of the relaxation kinetics of these subsystems can be treated independently, assuming that the energy of each of them is the quasi-integral of motion.

In the case of spin $I^z > 1/2$ the validity of the introduction of spin temperature is much less certain. However, if the interaction of the spins may be regarded as purely Zeeman, the energy levels are equidistant and the spin temperature can be introduced by the relation analogous to (6.49).

Of course, for any initial values of the population the mutual tilting of the spins lead to the establishment of quasi-equilibrium at time T_2, with the latter corresponding to the most probable distribution of the population in terms of constancy of total energy. If the energy is distributed between identical subsystems then, according to the principles of classical statistical thermodynamics, the most probable distribution by levels is the Boltzmann distribution.

So far we have discussed the simplest way of introducing the spin temperature theory. Non-equilibrium statistical thermodynamics suggests a more reasonable approach which can be used to consistently study the thermodynamics of nuclear spins in solids. In order to use the basic definitions of thermodynamics, we divide the full system with individual Zeeman levels into three subsystems: the thermostat $\{i = 1\}$, which is a perfect lattice, the Zeeman subsystem $\{i = 2\}$ and the spin–spin subsystem $\{i = 3\}$. The full Hamiltonian of the system can then be written as

$$H = H_1 + H_2 + H_3 + H_{int}.$$

Here

$$H_2 = -\omega_0 \sum_i I_i^z$$

is the operator of the Zeeman energy of the nuclei in a constant magnetic field $\tilde{H}$, $\omega_0 = \gamma \tilde{H}$ is the Larmor frequency,

$$H_3 = \frac{1}{2} \sum_{ij} \sum_{\alpha\beta} u_{ij}^{\alpha\beta} I_i^{\alpha} I_j^{\beta}$$

is the dipole – dipole interaction of nuclear spins [137]; H_{int} is the interaction Hamiltonian of nuclear spins with the lattice. The lattice Hamiltonian H_1 has the standard form.

If the amplitude of the external field is small and the probability of transitions induced by the external field is less than the transition probabilities related to the interaction within the system, the system as a whole may be in a quasi-equilibrium state [44,65]. Assuming that this condition is satisfied, we construct non-equilibrium and quasi-equilibrium operators using the method described in [44]. We get

$$\rho(t) = Q^{-1} \exp\left\{-\sum_j \beta_j H_j - \beta_1 (H_1 + H_{int}) + \right.$$

$$\left. + \sum_j \int_{-\infty}^0 d\tau e^{\epsilon t} (\beta_j - \beta_1) K_j(\tau)\right\}, \tag{6.50}$$

$$\rho_q = Q_q^{-1} \exp\left\{-\sum_j \beta_j H_j - \beta_1 (H_1 + H_{int})\right\}.$$

Here Q is the statistical functional, $\epsilon \to +0$, $j = (2, 3)$

$$K_j = \frac{1}{i}\left[H_j, H_{int}\right].$$

According to the definitions of non-equilibrium statistical thermodynamics, the parameters

$$\beta_2 = \frac{\partial S}{\partial \langle H_2 \rangle} \text{ and } \beta_3 = \frac{\partial S}{\partial \langle H_3 \rangle}$$

(S is entropy) have the meaning of the reciprocal spin quasi-temperature Zeeman reservoir {2} and the reciprocal quasi-temperature dipole–dipole reservoir {3}. In addition, in what follows we restrict our study to the case when the lattice specific heat is much greater than the heat capacity of the spin system. We can then assume that, regardless of whether there is an energy exchange between the spin system and the lattice or not, the thermostat is always in thermodynamic equilibrium with constant temperature $1/\beta_1$.

One more remark. In determining the reciprocal spin quasi-temperature the following average values were taken into account

$$\langle H_2 \rangle = \mathrm{Sp}(\rho H_2), \quad \langle H_3 \rangle = \mathrm{Sp}(\rho H_3).$$

Moreover, in accordance with the basic definitions of the non-equilibrium operator method, the parameters β_2 and β_3 are chosen so as to satisfy the thermodynamic equation

$$\langle H_j \rangle = \langle H_j \rangle_q = -\frac{\delta}{\delta \beta_j}(\ln Q_q),$$

where $Q_q = \mathrm{Sp} \exp\left\{-\sum_j \beta_j H_j - \beta_1 (H_1 + H_{int})\right\}$; $\langle H_j \rangle_q = \mathrm{Sp}(\rho_q H_j)$.

There are two possible physical situations that lead to a highly non-equilibrium state. The first of them will be considered. Suppose that initially (i) the Zeeman and spin–spin subsystems are in quasi-equilibrium state. Then, after a short time $t \ll T_2$ the external magnetic field changes from $\tilde{H}^{(i)}$ to $\tilde{H}^{(f)}$. After relaxation (in time $t \ll T_1$) system goes into a new quasi-equilibrium state with the spin quasi-temperature common for the subsystems {2} and {3}. In the traditional approach [137] we can show that the total reciprocal quasi-temperature $F_s^{(1)}$ in the final state (f) is equal to the average weighted value, which can be obtained if the parameters of β_2 and β_3 are taken with their statistical weights. As the weight factors are selected heat subsystem {2}, namely $\left(\tilde{H}^{(f)}\right)^2 \mathrm{Sp}(M_z^2)$, and heat capacity of the subsystem {3} $\mathrm{Sp}(H_3^2)$. (Here, M_z is magnetization). Under this approach, we get

$$F_s^{(1)} = \frac{\beta_2 \left(\tilde{H}^{(f)}\right)^2 \mathrm{Sp}(M_z^2) + \beta_3 \mathrm{Sp}(H_3^2)}{\left(\tilde{H}^{(f)}\right)^2 \mathrm{Sp}(M_z^2) + \mathrm{Sp}(H_3^2)}. \tag{6.51}$$

Next, we consider the second possible situation. Suppose that we place the spin subsystem also in an additional radio-frequency field $2\tilde{H}_1 \cos \omega t$. Consequently, the quasi-equilibrium state of this subsystem has moved. We explain this fact in more detail. Let the Zeeman levels be separated by energy intervals $\Delta E_s = \omega_0$. Then, the time-dependent perturbation, which is the RF field, will induce transitions between Zeeman levels, provided that the field frequency ω is close to the Larmor frequency ω_0. We will not study the resonance effects. Therefore, we assume that initially the magnetic field is removed from the resonant value by the value $\gamma_h = \omega - \omega_0$.

In the case when the RF field pumps energy into nuclear spins, the expression for the total spin quasi-temperature $1/F_s^{(2)}$ of two subsystems $\{2\}$ and $\{3\}$ has a complicated form. Therefore, we confine ourselves to studying a particular case. Let us assume that at the initial time (i), when the RF field does not yet exist, the quasi-temperatures of the subsystem $\{2\}$ and $\{3\}$ are equal to the lattice temperature $1/\beta_1$ as a result of spin–lattice relaxation. The RF field is then instantaneously (within fractions of a microsecond) applied to the system so that at the end of the pulse (f) the following values re obtained: $\tilde{H}_1 = \tilde{H}_1^{(f)}$ and $h = h^{(f)}$. As a result, in the non-equilibrium state, which is achieved under the influence of strong magnetic fields and radio-frequency field, the total spin quasi-temperature of the subsystems $\{2\}$ and $\{3\}$ takes the form

$$\frac{1}{F_s^{(2)}} = \frac{1}{\beta_1}\left\{1 + \frac{\left(\tilde{H}_1\right)^2}{\overline{h}^2\left(\tilde{H}_L\right)^2}\right\}. \tag{6.52}$$

Here $\overline{h}^2 = \overline{\left(h^{(i)}h^{(f)}\right)}$, where the bar denotes the algebraic averaging, and $\left(\tilde{H}_L\right)^2$ is the square of the internal local field, in other words, the Weiss field.

We make several observations concerning the expression (6.52):

– Quasi-temperature $1/F_s^{(2)}$ (6.52) is, of course, determined by the balance between pumping energy provided by the radio-frequency field and the flux of energy into the lattice through spin-lattice interaction. However, in this case, the stage of establishment of quasi-temperature proceeds in a very short time interval, more precisely, the interval that is less than the spin–lattice relaxation time T_1. So, to a first approximation the spin subsystems $\{2\}$ and $\{3\}$ can be considered 'isolated' from the lattice. The parameters of the spin–phonon interaction are not included in the formula (6.52).

– It is necessary to specify the particular format of the value $(\tilde{H}_L)^2$. Although the RF field depends on time, the evolution of the system can be described with good accuracy by the Hamiltonian which is almost completely independent of time. To do this, it would seem sufficient just to go to the coordinate system rotating around the direction of the magnetic field with frequency ω.

However, analysis shows that, strictly speaking, even with this formulation it is not possible to completely eliminate the time dependence. Therefore, when studying the process of establishing quasi-temperature $1/F_s^{(2)}$, initally the equation is derived without interaction with the lattice, and the resultant expression is corrected using a correction due to the spin–lattice interaction. Note also that in the calculation in the rotating coordinate system, formula (6.52) (with a small correction) retains its form, but $(\tilde{H}_L)^2$.in the expression for quasi-temperature is replaced by $(\tilde{H}'_L)^2$ – the square of the local internal field in the rotating system.

– We note some similarities in the form of writing the quasi-temperature in the theory of channeling (6.27) and spin quasi-temperature (6.52). The difference between the two formulations lies in the fact that (6.27) includes the contribution of the process of pumping energy and the contribution of the dissipative process. Then, as in (6.52), the effect of the dissipative process is not considered. It should therefore be expected that the formula (6.52) describes an overvalued 'warm up' of the spin subsystem.

Suppose that an external magnetic field is so strong that the spin-lattice relaxation time is much greater than the rate of exchange of heat between the subsystems {2} and {3}; exchange is due to non-secular terms of the spin–spin interaction. In this case, the average energies of the two spin subsystems change independently of each other, and the subsystems approach equilibrium with the thermostat at different speeds. Expressing the time derivatives of the average values of $\langle H_j \rangle_q$ in terms of derivatives of reciprocal quasi-temperatures, we find that the process of approaching equilibrium is described by the relaxation equation of the form

$$\frac{d}{dt}\beta_j(t) = -\frac{1}{\tau_j}\{\beta_j(t) - \beta_1\}, \qquad (6.53)$$

where

$$\tau_j = \langle H_j^2 \rangle_q \left(L_{jj}\right)^{-1}, \quad L_{jj} = \int_{-\infty}^{0} dt\, e^{\varepsilon t}\left(K_j, K_j(t)\right),$$

L_{jj} are the Onsager kinetic coefficients [23].

We assume that the interaction of spins with the lattice is described by the operator

$$H_{int} = \sum_{\alpha} V_{\alpha} f_{\alpha}(t).$$

Here V_{α} are the spin operators, and $f_{\alpha}(t)$ is a normal Markov process with zero mean and the exponential correlation function

$$\langle f_{\alpha}(t) f_{\alpha}(0) \rangle = B_{\alpha}(0) \exp\left\{ -\frac{t}{\tau_c} \right\}.$$

The correlation time τ_c, which became part of the correlation function, can be compared in particular cases with the Larmor precession period. Omitting the detailed calculations, we represent the final expressions of the reciprocal relaxation time for the Zeeman and spin–spin subsystems, which are contained in (6.53)

$$\frac{1}{\tau_1} = \frac{2\gamma^2 \bar{h}^2 \tau_c}{1 + \omega_0^2 \tau_c^2}, \quad \frac{1}{\tau_2} = \frac{2}{3} \gamma^2 \bar{h}^2 \tau_c \left(1 + \frac{5}{1 + \omega_0^2 \tau_c^2} \right).$$

As shown by the assessment, condition $\tau_1 \gg \tau_2$ is satisfied at $\omega_0 \tau_c \gg 1$ so that the dipole–dipole subsystem {3} is thermalized faster than Zeeman's {2}.

6.7.2. Statistical characteristics of channeled particles and the spin gas

It is interesting to compare the distribution of CPs $p(\varepsilon_\perp)$ (Fig. 6.4a, where curve 3 from Fig. 5.3 is repeated but on a different scale) with the distribution of the subsystem of the nuclear spins isolated from the lattice. This situation can be realized in an experiment with a lithium fluoride crystal with a long spin-lattice relaxation time. The isolated spin subsystem behaves like an ideal spin gas and its full probability distribution can be written as [60,138]

$$\tilde{p}(E_s) = \tilde{G}(E_s) \tilde{n}(E_s).$$

Here, the probability of finding spins with energy E_s is given by the canonical distribution $\tilde{n}(E_s) \sim \exp(-F_s E_s)$, and $\tilde{G}(E_s)$ is the density of states of isolated spins. The curve $\tilde{p}(E_s)$ in Fig. 6.4b illustrates a situation where the rapidly varying density function $\tilde{G}(E_s)$ in combination with decreasing function $\tilde{n}(E_s)$ gives a sharp peak in the full distribution $\tilde{p}(E_s)$.

Another note. Being slightly less rigorous, we hereafter use the term 'ideal' gas, meaning the Zeeman subsystem which does not interact with the lattice, with paramagnetic impurities, and that does not have all the electron–nuclear interactions. As for the spin–spin (dipole–dipole) interactions, it is impoirtant to keep the description of the spin subsystem, since it is necessary to establish a uniform (equal) quasi-temperature throughout the volume of the so-called ideal gas.

Note the following. Systems, which are usually considered in statistical thermodynamics, have an unlimited energy spectrum from the top. This is explained by the presence in the Hamiltonian of the term corresponding to kinetic energy. If the position of the levels in the system is defined, statistical mechanics provides a method

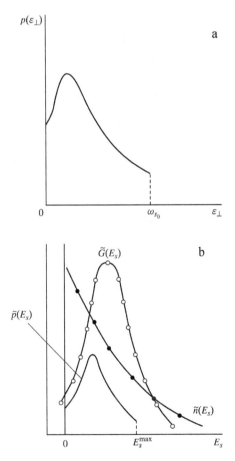

Fig. 6.4. The distribution of channeled particles with respect to transverse energy (a) and the energy distribution of the spin gas (b),

of finding the thermodynamic functions. It is sufficient to calculate the statistical sum, being the sum of the statistical weights of all levels of the system. If the energy spectrum extends to infinity, the temperature of the system must be positive, since otherwise the partition function would diverge. This situation is seen in the channeling task, where the levels of below-barrier states ($0 < \varepsilon_\perp < \omega_{s_0}$) borders at the top with the levels of above-barrier states. Therefore, the transverse quasi-temperature of the CPs is always positive.

As for the Hamiltonian of the spin subsystem, it does not contain the term corresponding to the kinetic energy, so that the energy spectrum is bounded from above. If the direction of the magnetic field is quickly reversed, the spin gas in which the lower levels were predominantly occupied, will be in a position with the largest population of the upper levels. In other words, a population inversion is observed, in which the spin temperature takes negative values. Further, the distribution of the population gradually changes, going initially through the first stage of the same population of all levels, which corresponds to an infinite spin temperature $1/F_s \to \infty$. The normal population, characteristic of thermodynamic equilibrium of the spins with the lattice, is then reached. It should be noted that in the region of negative spin temperatures the partition function of the spin subsystem is finite since the number of levels is limited.

The CP distribution $p(\varepsilon_\perp)$ is a non-Maxwellian function truncated, as shown in Fig. 6.4a, at high energies, $\varepsilon_\perp = \omega_{s_0}$. The distribution of an ideal spin gas $\tilde{p}(E_s)$ is also depicted (see Fig. 6.4b) by the non-Maxwellian function truncated at $E_s = E_s^{max}$. So that comparison of these distributions reveals their similarities in terms of statistical theory.

If we restrict ourselves to the range of positive quasi-temperatures, the shape of the distribution of population of the CP and the ideal spin gas have also much in common. In particular, they coincide in the limit of infinite quasi-temperature $(1/F_{21}) \to \infty$ and $(1/F_s) \to \infty$, when the uniform population of all levels of the CP subsystem is reached in the model of truncated oscillators and uniform population of the levels of the ideal spin gas. It is interesting to note that the uniform population is actually the starting point of all theories of channeling, developed within the framework of statistical mechanics [13,14,31,79,139]. All these works featured the density matrix for the microcanonical statistical ensemble, so the main independent thermodynamic variable – quasi-temperature of the CPs – in missing in these papers.

Thus, the truncated model and the oscillator model of an ideal spin system with the maximum energy $E_c^{\max}$ proved very useful for understanding the statistical and thermodynamic meaning of the physical quantities; studies based on these models revealed similarity of fast particles, moving in the channeling mode, and the ideal spin gas. It should be remembered that the physical situation, typical for the spin system, is possible only in the ideal gas model; the interaction with the lattice leads to a change in energy levels and a change in the average population of these levels.

If in the calculation of the spin quasi-temperature we take into consideration the contribution of the interaction of the spins with the lattice, then (6.52) includes an additional factor – the relative frequency of the spin-lattice relaxation $\eta_s = \dfrac{\tau_2}{\tau_3}$. More precisely, the energy $\left(\tilde{H}_L\right)^2$ in the denominator of the second term of (6.52) is replaced by the value $\left(\tilde{H}_L\right)^2 \eta_s$, which describes the interference of spin–spin new interaction with the weaker spin–lattice interaction.

7

Non-equilibrium thermodynamics of chaotic particles

7.1. Relaxation kinetics and quasi-temperature of positronium atoms

7.1.1. Ore model. The energy balance equation

Positronium is formed in collisions of slow positrons with atoms of matter, followed by the capture of an electron by positron from atoms of the medium. Positrons passing through a solid or gas lose some of their energy due to ionization of the medium. In addition to the ionization process, the formation of positronium atoms is significantly affected by the excitation of electronic levels of atoms in a solid. Moreover, there is a significant decline in the probability of positronium formation, when the positron energy exceeds the excitation energy of the first electronic level $\overset{*}{E}$. Recall that the energy $\overset{*}{E}$ is always lower than the ionization potential V_{ion}.

Full analysis of the probability of positronium formation after slowing down of the positrons moving in a solid, to the energy of several tens of electron volts requires consideration of the contributions of many elementary processes. Namely, it is important to consider the cross section of ionization and excitation of lattice atoms under the influence of the positrons, elastic and inelastic scattering of positrons by the lattice atoms, the capture cross section of the electron by the positron and positronium dissociation in a collision with a lattice atom. Since the fraction of positrons that form positronium atoms, taking into account all of the elementary processes, cannot be calculated, it is necessary to use a simpler model, the so-called Ore model [139]. In this model, the energy range of positrons in which the formation of a hydrogen-related electron-positron system is most likely, is determined. Given the fact that the binding energy of the positron and the electron is equal to half the ionization potential of the hydrogen atom [140–144]

$$E_B = \frac{1}{2} \text{Ry} = 6.8 \text{ eV}, \tag{7.1}$$

the lower boundary of the Ore gap is $E_{min} = V_{ion} - E_B$. The upper limit has the form of the ionization potential of the atoms in the lattice, so that in case of formation of positronium in solids we have $E_{max} = V_{ion}$.

We assume that the positrons, moving in a solid, slowed to E_{max}. If we assume that the energy of the positrons in the range from zero to E_{max} is uniformly distributed, then all the positrons with energies within the Ore gap form positronium with equal probability. According to [139], this probability is $P = (\Delta/E_{max})$, where Δ is the Ore gap width.

A more accurate expression for the probability of positronium formation can be obtained by using two sections: σ_{p_s} – section of positronium formation, and σ_t – the total cross section of all processes leading to slower motion of positrons. Note that the ratio $\gamma = \sigma_{p_s}/\sigma_t$ is usually much less than unity and is independent of the kinetic energy of the positron, E. Under this approach, we can easily obtain the probability of formation, averaged over the energies inside the gap. We have

$$\overline{P} = \begin{cases} \dfrac{\Delta}{E_{max}} & \text{at } \gamma \gg \xi, \\[2em] \dfrac{\gamma}{\xi} & \text{at } \gamma \ll \xi, \end{cases}$$

where ξ is the average energy lost in a single collision.

If the probability $P(E)$ is not averagred over energies, we get quite cumbersome expressions. The energy dependence described by these expressions is shown schematically in Fig. 7.1 [139]. An increase of $P(E)$ to the value of $P(\overset{*}{E})$ in the figure is due to the fact that as the variable E approaches $\overset{*}{E}$ from the side of lower energies, the number of collisions that slow down the movement of a particle increases. At the same time, decrease of $P(E)$, observed in the region

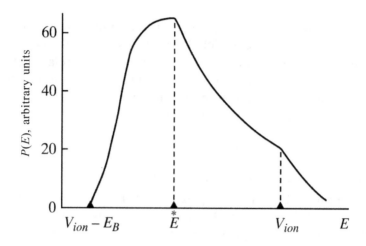

Fig. 7.1. Probability of formation of a positronium atom as a function of the kinetic energy of the positron.

$\overset{*}{E} < E < V_{ion}$, is associated with the processes of excitation of electronic levels, and an additional decline at $E > V_{ion}$ – with ionization.

Assuming that the positronium atoms (PA) are an independent thermodynamic subsystem, we divide the full system into two weakly interacting subsystems: the index $i = 1$ corresponds to the thermostat (lattice + electron gas), the index $i = 2$ corresponds to the gas of 'hot' PA. In the reduced description the state of the system is characterized by a set of average values of two variables P_{mi}, where $m = 1, 2$, namely, $P_{1i} = H_i$ – Hamiltonian, and $P_{2i} = N_i$ – number of particles of the i-th subsystem. The total Hamiltonian of the PA has the form $H_2 = H_0^{(2)} + H_{int}^{(2)}$, where $H_0^{(2)}$ is the Hamiltonian of the ideal gas of the PA, $H_{int}^{(2)}$ is the Hamiltonian of interaction of PA with the thermostat. The interaction between the PA is not considered, since the number density of particles in the subsystem (2) is sufficiently small.

According to [44], the entropy functional can be written as

$$S(t,0) = \Phi(t) + \sum_{mi} P_{mi} F_{im}(t),$$

where

$$\Phi(t) = \ln \mathrm{Sp} \exp \left\{ -\sum_{mi} P_{mi} F_{im}(t) \right\}, \quad F_{im}(t) = \frac{\delta S(t)}{\delta \langle P_{mi} \rangle^t}.$$

We represent the non-equilibrium statistical functional as an expansion in powers of the interaction potential. Then up to the first order we have

$$\rho(t) = \rho_I(t) + \int_{-\infty}^0 dt' e^{\epsilon t'} \int_0^1 d\tau\, e^{-\tau S_0(t,0)} \Delta\left\{ \left[F_{21}(t+t') - \beta_1 \right] \dot{H}_2(t') \right\} \times$$
$$\times e^{(\tau-1)S_0(t,0)} - \int_0^1 d\tau\, e^{-\tau S_0(t,0)} \left\{ \beta_1 H_{int}^{(2)} \right\} e^{(\tau-1) S_0(t,0)}. \tag{7.2}$$

Here $\rho_I(t) = \exp\{-S(t,0)\}$ is the quasi-equilibrium distribution corresponding to an extremum of the information entropy,

$$S_0(t,0) = \Phi_0(t) + F_{21}(t)\{H_2 - \tilde{\mu}_2 N_2\} + \tilde{B},$$
$$\Phi_0(t) = \ln \text{Sp} \exp\left\{ -F_{21}(t)(H_2 - \tilde{\mu}_2 N_2) - \tilde{B} \right\}, \tag{7.3}$$
$$\tilde{B} = \beta_1 (H_1 - \tilde{\mu}_1 N_1),\ \tilde{\mu}_2 = \mu_2 - \epsilon,\ \Delta A = A - \langle A \rangle^t,\ \epsilon \to +0.$$

Parameters $F_{im}(t)$, included in (7.2) and (7.3), are chosen so that the true average set of values $\langle P_{mi} \rangle^t = \text{Sp}\{\rho(t) P_{mi}\}$ is equal to their average quasi-equilibrium $\langle P_{mi} \rangle_I^t = \text{Sp}\{\rho_I(t) P_{mi}\}$. Then the quasi-equilibrium distribution is close to the true non-equilibrium distribution. With this choice of $F_{im}(t)$ the thermodynamic parameters are: $1/F_{11} = 1/\beta_1 = T_1$ – thermostat temperature, $1/F_{21}(t)$ – quasi-temperature of subsystem $\{2\}$, $F_{12} = -\beta_1 \mu_1$, μ_1 – chemical potential of the thermostat, $F_{22}(t) = -F_{21}(t)\mu_2$, $\mu_2 = \mu_p + \mu_e$ – chemical potential of the subsystem $\{2\}$, $\lambda = \exp\{-F_{22}(t)\}$ – the absolute activity of the PA; $\epsilon = (1/\overline{N}_2)\langle H_2^0 \rangle_0$, where the averaging is performed using statistical operator $\rho_0 = \exp\{-S_0(0,0)\}$, $\overline{N}_2$ – the number of PA. Using the expression for the Massieu–Planck functional $\Phi(t)$, it is easy to derive relationships describing the fluctuations of the quantities P_{mi},

$$\frac{\delta^2 \Phi(t)}{\delta F_{im}(t) \delta F_{jn}(t)} = -\frac{\delta \langle P_{nj} \rangle^t}{\delta F_{im}(t)} = -\frac{\delta \langle P_{mi} \rangle^t}{\delta F_{jn}(t)} = (P_{nj}, P_{mi})^t, \tag{7.4}$$

where

$$(P_{nj}, P_{mi})^t = \int_0^1 d\tau\, \text{Sp}\left\{ P_{nj} e^{-\tau S_0(t,0)} \Delta P_{mi} e^{(\tau-1) S_0(t,0)} \right\}. \tag{7.5}$$

Using the thermodynamic relations (7.4) and (7.5), we establish the connection of the quasi-temperature subsystem $\{2\}$ with energy fluctuations at the final temperature of the thermostat. We get

$$\frac{1}{F_{21}(t)} = \left\{ \frac{1}{C_V} (H_2, H_2)' \right\}^{1/2}, \tag{7.6}$$

where C_V is the specific heat of the PA gas. In the steady quasi-equilibrium state the relation (7.6) is greatly simplified. In this case, (the steady state) using (7.5) and (7.6), we find

$$\left(\frac{1}{F_{21}} \right)_{ss} \equiv \frac{1}{\beta_2} = \left\{ \frac{1}{C_V} (H_2, H_2) \right\}^{1/2}, \tag{7.7}$$

where

$$(H_2, H_2) = \int_0^1 d\tau \operatorname{Sp} \left\{ H_2 e^{-\tau S_0(0,0)} \Delta H_2 e^{(\tau-1) S_0(0,0)} \right\}.$$

To construct the energy balance equation of the PA in the quasi-equilibrium state (or close to it), it is necessary to calculate, on the one hand, the power of energy input to the subsystem {2}, due to external radiation, on the other hand the power of the energy transferred to the thermostat. The average rate of increase of the internal energy density of the PA $E_2(t) = \langle H_2 \rangle^t$ in the case of irradiation of a dielectric is

$$\frac{1}{\overline{N}_2} \langle H_2 \rangle^t \left(\tilde{N} \tilde{\Phi} \sigma_f \right).$$

Here $\tilde{\Phi}$ is the flux density of bombarding positrons; σ_f is the cross-section of collisions of positrons with electrons (the number of electrons of the atomic cores); $\tilde{N}$ is the density of atoms in the lattice; $\tilde{N} \tilde{\Phi} \sigma_f$ is the rate of formation of the PA in unit volume. For simplicity, the volume of the system is assumed to be unity. In the transition from the expression written for the PA in a dielectric, the expression for the PA in a metal $\tilde{N}$ should be replaced by the density of conduction electrons.

In the steady quasi-equilibrium state, the rate of increase of internal energy can be written in expanded form using the expression [145]

$$\frac{1}{\overline{N}_2} \langle H_2 \rangle = \overline{\varepsilon} f_1 (F_{21}),$$

$$f_1(F_{21}) = 2 \left[\varphi(F_{21}) \right]^{-2} \left\{ 1 - \left[1 + \varphi(F_{21}) \right] \exp \left[-\varphi(F_{21}) \right] \right\}, \tag{7.8}$$

which is applicable to any subsystem of chaotic particles, including those for PA at the bottom of the zone. In the formula (7.8) $\varphi(F_{21}) = \frac{\pi}{2} F_{21} \bar{\varepsilon}$, and the average energy $\bar{\varepsilon}$, as in statistical mechanistic theory is given in the limit $1/F_{21} \to \infty$ [107,128]. Using (7.8), the energy balance equation that describes the energy situation in the subsystem {2}, has the form

$$\left\langle \frac{dE_2}{dt} \right\rangle = \bar{\varepsilon} f_1(F_{21}) \tilde{N} \tilde{\Phi} \sigma_f - L_{\dot{H}_2 \dot{H}_2} X. \tag{7.9}$$

Here $L_{\dot{H}_2 \dot{H}_2}$ is the transfer coefficient in a linear relation between the energy flux and the thermodynamic force $X = \beta_1 - F_{21}$, and $\bar{\varepsilon} = \lim \varepsilon$ at $1/F_{21} \to \infty$ in full compliance with [128]. The second term on the right side of (7.9) is the energy the subsystem {2} loses as a result of two dissipative processes: phonon scattering of PA and excitement of electron–hole pairs.

We note one more circumstance. With increasing penetration depth the correlation of the direction of the velocity of the particles with the direction of entry of the positrons in the crystal noticeably weakens due to multiple scattering. If the characteristic value is the depth at which positronium atoms form, then at greater depth the positronium atoms and free positrons can be considered as particles of the chaotic beam. Therefore, hereafter the equations are written only for chaotic particles.

7.1.2. Phonon and electron scattering on positronium atoms

We consider the intraband scattering of the PA $(1s, k) \to (1s, k + q)$ by thermal vibrations of lattice atoms. In this case, the optical mode of the vibration makes a small contribution to the kinetic coefficients [117]. Therefore, we restrict ourselves to the interaction of PA with acoustic phonons. In the region of small momentum $\mathbf{q}$ transfers the transfer coefficient, included in the balance equation (7.9), can be written in the form

$$L_{\dot{H}_2 \dot{H}_2}^{(ph)} = \frac{1}{2} \int d\omega \, \omega^2 \frac{1}{2\pi^2} \int d\mathbf{q} \left| V_1(q) \right|^2 \beta_1 \omega \times$$
$$\times S^{(ph)}(\mathbf{q}, \omega) \delta((\mathbf{q}\mathbf{v}(\bar{\varepsilon})) - \omega). \tag{7.10}$$

In (7.10) we use the notation: $V_1(q)$ – the Fourier component of the deformation potential of interaction of the PA with the lattice [146]

$$\left|V_1(q)\right|^2 = g^2 \frac{q}{2M\tilde{N}_{v_s}},$$

g is the coupling constant with the order of magnitude equal to the energy of the positron–vacancy complex; M and v_s are the mass of the atom of the lattice and the speed of sound;

$$S^{(ph)}(\mathbf{q},\omega) = \frac{1}{2\pi} \int_0^\infty dt \exp(i\omega t) \left\langle \rho_{-\mathbf{q}}^{(i)} \rho_{\mathbf{q}}^{(i)}(t) \right\rangle_0$$

is the dynamic form factor,

$$\left\langle \rho_{-\mathbf{q}}^{(i)} \rho_{\mathbf{q}}^{(i)}(t) \right\rangle_0 = \exp(-2\tilde{W}) \sum_{\mathbf{R}} \exp(-i\mathbf{q}\mathbf{R}) \times$$
$$\times \exp\left\langle \left(\mathbf{q}\mathbf{u}(0,t)\right)\left(\mathbf{q}\mathbf{u}(\mathbf{R},t)\right)\right\rangle_0,$$

where $\rho_{\mathbf{q}}^{(i)}$ is the Fourier component of the density of atoms in the lattice; $\mathbf{u}(\mathbf{R}, t)$ is the displacement of an atom from its equilibrium position $\mathbf{R}$; $\exp(-2\tilde{W})$ is the Debye–Waller factor. Integrating (7.10) over the angles of the vector $\mathbf{q}$, we have

$$L_{\dot{H}_2\dot{H}_2}^{(ph)} = \frac{1}{2\pi^2} \nu(\bar{\varepsilon})\beta_1 \int d\omega\, \omega^2 \int dq\, q^3 \left|V_1(q)\right|^2 \times$$
$$\times S^{(ph)}(q,\omega) \left| \int_0^\infty dt'\, e^{-i\omega t'} j_1\left(q\nu(\bar{\varepsilon})t'\right)\right|,$$

(7.11)

where $j_1(x)$ is the cylindrical Bessel function.

Now in (7.11), integrating over dt' and $d\omega$, the dynamic form factor is expanded in a series in powers of displacement. If we confine ourselves to one-phonon and two-phonon scattering processes of the PA, then it is sufficient to keep terms of order u and u_2 in (7.11). The transport coefficient (7.11) is then written as

$$L_{\dot{H}_2\dot{H}_2}^{(ph)} = L_{\dot{H}_2\dot{H}_2}^{(1)} + L_{\dot{H}_2\dot{H}_2}^{(2)},$$
$$L_{\dot{H}_2\dot{H}_2}^{(1)} = \bar{\varepsilon}\, T_1 \frac{1}{\tau_1}, \quad L_{\dot{H}_2\dot{H}_2}^{(2)} = \bar{\varepsilon}\, T_1 \frac{1}{\tau_2}.$$

(7.12)

In the Debye model, the relaxation times τ_1 and τ_2, corresponding to one-phonon and two-phonon scattering, are

$$\frac{1}{\tau_1} = \frac{1}{3\pi^2} \frac{1}{M\tilde{N}} \int_0^{qD} dq\, q^6 \frac{1}{2M\omega_q} \left|V_1(q)\right|^2 \beta_1^2 v\left(\omega_q\right),$$

(7.13)

$$\frac{1}{\tau_2} = \frac{4}{3\pi^2} \frac{1}{M\tilde{N}} \int_0^{qD} dq\, q^6 \left[\frac{1}{\omega_q^3}\left|V_1(q)\right|^2\right]^2 \times$$

$$\times \frac{1}{M} \int_0^{\Delta q} d\tilde{\omega}\,\tilde{\omega}\left(\Delta_q - \tilde{\omega}\right)\beta_1^2 v(\tilde{\omega}) v\left(\Delta_q - \tilde{\omega}\right),$$

(7.14)

$$v(\omega) = \left[\exp\left(\beta_1\omega\right) - 1\right]^{-1}.$$

In (7.13) and (7.14): q_D – Debye momentum; Δq – the energy transferred to the PA during the absorption (emission) of two phonons. Note that in specific experiments, even at liquid helium temperatures, the main contribution to the dissipative processes (V-contribution) comes from phonon scattering, if the PA is moving in a dielectric.

Consider the scattering of the PA with an energy corresponding to the bottom of the Ore gap (~5 eV) for conduction electrons. It should be noted that the plasmon energy [147] in the alkali metals is much higher than in the PA and in these circumstances the transfer of energy from the subsystem {2} to the thermostat through the creation of plasmons is impossible. (The same statement was made in [148]). The physical situation considered here is completely different than in the passage of high-energy positrons (~1 keV), when the energy dissipation is caused by the plasmons. We assume that the removal of energy is completely determined by the processes of excitation of electron–hole pairs near the Fermi surface.

As noted in Section 7.1, the PA can be considered as a hydrogen atom [146]. This simplified model is adopted in our work. So the potential of interaction of electrons with the PA will be represented by the potential, averaged over the ground state of the intrinsic electron ψ_0,

$$V_2(q) = \frac{4\pi e^2}{q^2}\left\{1 - \int dr\, \frac{4\pi}{q} r \sin(qr)\left|\psi_0(r)\right|^2\right\},$$

(7.15)

where $\left|\psi_0(r)\right|^2 = \pi^{-1} a_B^{-3} \exp\left(-2r/a_B\right)$, a_B – the Bohr radius.

The general form of the transfer coefficient in the elastic intraband scattering with small momentum transfer is known [123]

$$L^{(e)}_{\dot{H}_2 \dot{H}_2} = \frac{1}{2} \int_{-\infty}^{\infty} d\omega\, \omega^2 \int d\mathbf{q}\, P(\mathbf{q}, \omega),$$

(7.16)

where

$$P(\mathbf{q},\omega)\,d\mathbf{q}\,d\omega =$$

$$= -\frac{1}{(2\pi)^2}\left|V_2(q)\right|^2\left|S^{(e)}(\mathbf{q},\omega)\delta((\mathbf{q}\mathbf{v}(\bar{\varepsilon})))-\omega\right)d\mathbf{q}\,d\omega \qquad (7.17)$$

is the transition probability, and the dynamic form factor is expressed in terms of the Fourier component (at time 'variable) of Green's function

$$G(\mathbf{q},t) \equiv \left\langle\!\left\langle \rho_q(t)_{\rho_q}^+ \right\rangle\!\right\rangle = -i\theta(t)\left\langle\!\left[\rho_q(t),\rho_q^+(0)\right]\!\right\rangle_0,$$

$$S^{(e)}(\mathbf{q},\omega) = \frac{1}{\pi}\,\mathrm{Im}\left\langle\!\left\langle \rho_q\big|\rho_q^+\right\rangle\!\right\rangle_\omega,$$

where ρ_q is the Fourier component of the electron density fluctuations.

Equation (7.16) will be transformed. First, confining ourselves to the electron gas at zero temperature, we take out from under the integral sign with respect to $d\omega$ factor ω at the point $\omega = (3/5)\varepsilon_F$, corresponding to the average energy of the electron in the ground state. (The above approximation provides sufficient accuracy [123]). Second, we carry out in (7.16) and (7.17) integration over the angular variables $\mathbf{q}$ and we use the approximation of $j_1(qvt) \approx 1/3qvt$ which is applicable at low velocities. In conclusion, in the expression obtained in this way we perform integration over $d\omega$, using the theorem of residues. The result is

$$L^{(e)}_{\dot{H}_2\dot{H}_2} =$$

$$= -\frac{1}{4\pi^4}\varepsilon_F v^2(\bar{\varepsilon})\int dq\,q^4\left|V_2(q)\right|^2\left\{\frac{d}{d\omega}\,\mathrm{Im}\,G(q,\omega)\right\}_{\omega=0}. \qquad (7.18)$$

Calculating the derivative of Green's function, performed in the Hartree approximation [149], leads to the result

$$\frac{d}{d\omega}\,\mathrm{Im}\,G(q,\omega)\bigg|_{\omega=0} =$$

$$= \begin{cases} -(2qv_F)^{-1}\,\mathcal{N}(\varepsilon_F)\left[1+\dfrac{k_{\mathrm{TF}}^2}{q^2}f_2(q)\right]^{-2} & \text{at } 0<q<2k_F \\[4mm] 0 \text{ at } q>2k_F. \end{cases} \qquad (7.19)$$

In the formulas (7.18), (7.19): $k_F = m_e v_F$; ε_F is Fermi energy; m_e is the mass of the electron; k_{TF} is the Thomas–Fermi reciprocal screening radius; $\mathcal{N}(\varepsilon_F)$ is the density of electron states per unit energy near the Fermi surface

$$f_2(q) = 1 - q^2 \left[2\left(q^2 + k_F^2\right)\right]^{-1}.$$

Now we introduce in (7.11) and (7.12) a new variable $\xi = (q/2k_F)^2$. We have

$$L_{\dot{H}_2\dot{H}_2}^{(e)} = \overline{\varepsilon}\,\varepsilon_F \frac{1}{\tau_3}, \tag{7.20}$$

where

$$\frac{1}{\tau_3} = \frac{2}{\pi^4}\varepsilon_F \int_{\xi_{min}}^{1} d\xi\,\xi^3\, \frac{1}{\left(\xi+\gamma\right)\left(\xi+\alpha\,f_2(\xi)\right)^2}. \tag{7.21}$$

Here we use the notation:

$$\alpha = \frac{e^2}{\pi v_F} = \left(\frac{k_{TF}}{2k_F}\right)^2, \qquad \gamma = \left(2k_F a_B\right)^{-2}.$$

The integration over $d\xi$ is limited by the value $\xi_{min} = (\omega p/\varepsilon_F)^2$, ω_p is the plasmon energy, because it does not take into account the contribution of the collective mode. We make two additional observations. Firstly, both transport coefficients (7.12) and (7.20) are proportional to the energy of PA $\overline{\varepsilon}$, which is typical for the transport coefficients at low energies. Second, the function $f_2(\xi)$ in (7.21) characterizes the contribution of correlation effects to the kinetic coefficient, calculated in the Hartree approximation.

7.1.3. Eigenvalue of the relaxation matrix and quasi-temperature of positronium

Using (7.9), the stochastic equation for the fluctuations of the internal energy $\delta E_2(t)$ can be written as

$$-\frac{d}{dt}\delta E_2(t) + \overline{\varepsilon}\left(\tilde{N}\tilde{\Phi}\sigma_f\right)\delta f_1(F_{21}) = F_{21}^2 L_{\dot{H}_2\dot{H}_2}\delta\left(\frac{1}{F_{21}}\right) - \zeta(t). \tag{7.22}$$

For simplicity, we restrict ourselves to the two-phonon processes, since they are most effective at high temperatures that are of interest in this problem. Therefore, in the case of PAs in the dielectric the kinetic coefficient (7.22) actually means $L_{\dot{H}_2\dot{H}_2}^{(2)}$.

Note two things. First, according to the Onsager hypothesis, the fluctuation $\delta A(t) = A(t) - \langle A(t)\rangle^2$ is analogous to the macroscopic deviation of the physical quantity and differs from it only by the fact that fluctuation occurs spontaneously. Because of this fluctuation $\delta E_2(t)$ satisfies the same form as the equation for the average, but with the additional term $\zeta(t)$, which should be considered as a random Gaussian variable.

Second, the state of the gas of the PA approaches quasi-equilibrium in a short time (about a picosecond). In the time scale of heat transfer used to study the transition to quasi-equilibrium, the details of the formation of the PA gas do not play any significant role. Therefore, the equation for the number of particles is excluded from consideration, but the equation (7.22) for energy fluctuations remains unchaged. The same formulation of the problem was used in many other studies, such as the theory of cascade processes [146], the theory of chemical reactions [23].

According to the results of non-equilibrium statistical thermodynamics [23], the covariance of a random term is proportional to the product $\eta = (\beta_2/\beta_1)^2$ and the transfer coefficient $L_{\dot{H}_2 \dot{H}_2}$. If, moreover, the effective mass of the particles is also taken into account, then there is an additional factor involving the square of the effective mass. We introduce an effective mass $\overset{*}{m}$ of the PA ($\overset{*}{m}$ definition is given below). Then the expression for the covariance of the random term is written in the form

$$\left\langle \zeta(t)\zeta(t')\right\rangle_{AV} = 2\left(\frac{\overset{*}{m}}{2m_e}\right)^2 \eta\, L_{\dot{H}_2 \dot{H}_2}\, \delta(t-t'), \qquad (7.23)$$

where $\left\langle ... \right\rangle_{AV}$ are the two-time averages used in the stochastic theory.

It is interesting to note that (7.23) describes the basic principle of the fluctuation–dissipation theory: fluctuations and the averages are determined by the same kinds of interactions.

Using the well-known formula $\delta(1/\beta_2) = \dfrac{1}{C_V}\delta E_2$, equation (7.22) can be written in the form of the relaxation equation

$$\frac{d}{dt}\delta E_2(t) = h\delta E_2(t) + \zeta(t).$$

The value of h, included in the relaxation equation, is a matrix in a general case. However, if the fluctuations of only one variable (energy) are considered, the matrix degenerates to a scalar h, which is the only eigenvalue of relaxation. Given that

$$\frac{\delta f_1(F_{21})}{\delta\left(\dfrac{1}{F_{21}}\right)} = \frac{8}{\pi}\left[\overline{\varepsilon}\,\varphi(F_{21})\right]^{-1} \text{ at } F_{21}\overline{\varepsilon} \gg 1,$$

we compare (7.22) with the relaxation equation. From the comparison it should be

$$h = \frac{1}{C_V}\left\{\frac{8}{\pi}\frac{1}{\varphi(F_{21})}\left(\tilde{N}\tilde{\Phi}\sigma_f\right) + F_{21}^2 L_{\dot{H}_2\dot{H}_2}\right\}. \qquad (7.24)$$

In the steady state, the PA quasi-temperature is expressed through the stationary energy – energy correlation function (7.7). The latter, in turn, determines the dispersion of the internal energy in the stationary quasi-equilibrium state (steady state)

$$\sigma_2^{ss} = (H_2, H_2)_0.$$

Given the type of covariance (7.23), we write the fluctuation-dissipation theorem, taking into account that h is a scalar, and we have [23]

$$2\sigma_2^{ss}h = -2\left(\frac{\overset{*}{m}}{2m_e}\right)^2 \eta L_{\dot{H}_2\dot{H}_2}.$$

We solve the resulting equation for dispersion σ_2^{ss}. As a result, taking into account the form of the relation matrix (7.24), we find

$$\sigma_2^{ss} = \frac{\left(\overset{*}{m}/2m_e\right)^2 \eta L_{\dot{H}_2\dot{H}_2}}{\dfrac{1}{C_V}\left\{F_{21}^2 L_{\dot{H}_2\dot{H}_2} - \dfrac{8}{\pi}\dfrac{1}{\varphi(F_{21})}\left(\tilde{N}\tilde{\Phi}\sigma_f\right)\right\}}. \qquad (7.25)$$

Now, with (7.7) and (7.25), we obtain the equation

$$\left(\frac{1}{\beta_2}\right)^2 = \frac{(\beta_2/\beta_1)^2\left(\overset{*}{m}/2m_e\right)^2 L_{\dot{H}_2\dot{H}_2}}{(\beta_2)^2 L_{\dot{H}_2\dot{H}_2} - \dfrac{8}{\pi}(1/\varphi(\beta_2))\left(\tilde{N}\tilde{\Phi}\sigma_f\right)}.$$

The solution of equation (7.26) in the case of the vanishingly small positron flux $\tilde{\Phi} \to 0$ (zero approximation) has the form

$1/\beta_2 = T_1\left(\overset{*}{m}/2m_e\right)$. Substituting the zeroth approximation in (7.26), in first approximation, we find

$$\frac{1}{\beta_2} = T_1 \frac{\left(\overset{*}{m}/2m_e\right)}{\sqrt{1 - T_1^2\left(\overset{*}{m}/2m_e\right)^3 \left(\tilde{N}\tilde{\Phi}\sigma_f\right)B_i}}. \tag{7.27}$$

Note that the approximate solution (7.27) is valid for the values of $\tilde{\Phi}$, which satisfy

$$\left(\tilde{N}\tilde{\Phi}\sigma_f\right) < \beta_1^2 \left(\frac{\overset{*}{m}}{2m_e}\right)^{-3} \frac{1}{B_i}.$$

The value of B_i in (7.27) takes two values, which are easily obtained by using (7.12) and (7.20)

$$B_i = \begin{cases} \left(\dfrac{4}{\pi\varepsilon}\right)^2 \tau_2 & \text{at } i = 2 \text{ (two-phonon scattering)} \\[2ex] \left(\dfrac{8}{\pi\varepsilon\,\varepsilon_F}\right)\dfrac{1}{\varphi(\beta_1)}\tau_3 & \text{at } i = 3 \text{ (electron scattering)} \end{cases} \tag{7.28}$$

The final expressions for the relaxation times of (7.28) can be obtained by continuing the integration in (7.14) (7.21). Find

$$\frac{1}{\tau_2} = \frac{4}{3\pi^2}\left(\frac{g}{Mv_s^2}\right)^4 \frac{1}{4\tilde{N}} \int_0^{q_D} dq\, q^2 \Delta_q, \tag{7.29}$$

$$\frac{1}{\tau_3} = \left(1 + \frac{3}{5}\alpha\right)^{-2}\left[\frac{1}{3}\left(1 - \xi_{\min}^3\right) - \right.$$
$$\left. -\frac{1}{2}\gamma\left(1 - \xi_{\min}^2\right) + \gamma^2\left(1 - \xi_{\min}\right) - \gamma^3 \ln\left|\frac{1+\gamma}{\xi_{\min} + \gamma}\right|\right]. \tag{7.30}$$

According to the criterion of non-equilibrium statistical thermodynamics [23], the steady state, far away from equilibrium, is thermodynamically stable if all eigenvalues of the relaxation matrix have negative real parts. In our case, there is only one eigenvalue (7.24), and the stability condition takes the simple form $h < 0$. This condition limits the range of permissible values of the quasi-temperature of the PA (7.27) for a given rate of formation of the PA

$$\left(\frac{1}{\beta_2}\right) < \left[\frac{\pi^2}{16}\bar{\varepsilon}L_{\dot{H}_2\dot{H}_2}\frac{1}{\tilde{N}\tilde{\Phi}\sigma_f}\right]^{1/3}.$$

Expression (7.27) describes the quasi-temperature of the PA $(1/F_{21})_{ss} \equiv 1/\beta_2$ as a renormalized thermostat temperature T_1. According to (7.27), the renormalization is associated with three factors. The first depends on the rate of formation of PA gas in a solid $\tilde{N}\tilde{\Phi}\sigma_f$. The second factor is due to kinetic effects of interaction. In order to present it explicitly, we write the real part of the mass operator corresponding to the bottom of the PA zone

$$\operatorname{Re}\tilde{M}(\bar{\varepsilon}) = \varkappa_1\bar{\varepsilon} + \varkappa_2\frac{\tilde{k}^2}{4m_e} + \Delta\mu_2.$$

Here $\Delta\mu_2$ is the shift of the chemical potential of the subsystem $\{2\}$; $\tilde{k}$ is the momentum of the PA. Since the effective mass expressed in terms of $\varkappa_1$ and $\varkappa_2$

$$\frac{m^*}{2m_e} = \frac{1-\varkappa_1}{1+\varkappa_2},$$

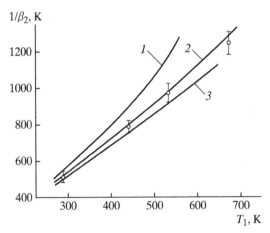

Fig. 7.2. Dependence of the quasi-temperature of a beam of positronium atoms on thermostat temperature.

the contribution of interaction effects in (7.27) is determined by the multipliers $m^*/2m_e$. Finally, the third renormalization factor is determined by dynamic effects of collisions and is described by means of the relaxation times τ_2 and τ_3 (7.29), (7.30), included in B_i (7.28).

Formula (7.27) gives a family of curves, some of which are shown in Fig. 7.2. (Curve 1 corresponds to $(\tilde{N}\tilde{\Phi}\sigma_f) B_2 = 65$ eV^{-2}, curve 2 – 27 eV^{-2}, curve 3 – the value of 8 eV^{-2}). As shown in the figure, the theoretical curve 2 is in satisfactory agreement with the experimental data obtained by analyzing the narrow component of the momentum distribution of the PA in quartz [117]. The quasi-temperature of the PA greatly deviates from the linear function with increasing thermostat temperature.

7.1.4. Thermalization of positronium atoms

There are two areas in the study of thermalization of fast particles in solids: a phenomenological approach [150], based on the heat conduction equation, and our proposed approach, in which the closed system is described on the thermodynamic level.

The difficulty of the phenomenological theory is not so much in finding the solution of the heat conductivity equation, but in the applicability of the concept of the theory of heat conduction. The fact is that in [115,150] the sources of the thermal spike are assumed to be cascade particles, positrons, etc. The energy that they convey to the lattice is released in a small volume of the medium. Is it possible in this case to assume that the linear heat conductivity equation, which describes the propagation of heat in an infinite, continuous medium, is applicable in these circumstances? In addition, the study of the cooling process in the conditions of a large temperature difference of the two subsystems is bound to lead to a quasi-linear heat conductivity equation, and not to the linear one [150].

The standard solution of the heat conductivity equation consists of two equal factors: a multiplier decreasing with increasing time and a multiplier that increases with time. Apparently, the desire to achieve a significant decrease of temperature is dictated by the fact that the authors of [150] completely reject the increasing mltiplier. In this case, the temporal evolution of the temperature of 'hot' atoms is determined by a multiplier of $\sim t^{-3/2}$, which has a singularity at $t = 0$. To avoid this difficulty, it is proposed to introduce a starting time t_0, using as a solution the function $\sim(t + t_0)^{-3/2}$, or to restrict the lower limit of integration in all integral transforms. Of course, all this adds additional uncertainty to the theory.

Despite the establishment of quasi-equilibrium in the PA subsystem with quasi-temperature (7.27), the overall system remains far from complete thermodynamic equilibrium, so the next stage in the evolution of the system is thermalization of the PA. At this stage, the PA subsystem passes through a sequence of states of this quasi-equilibrium. These states are described by the distribution $\rho_l(t) = \exp\{-S(t, 0)\}$, which includes the thermodynamic parameters $F_{im}(t)$, depending on the current time. The time evolution of the parameters $F_{im}(t)$ is described by a system of equations [44]

$$\frac{d}{dt}F_{im}(t) = \sum_{nj}\frac{\delta^2 S(t)}{\delta\langle P_{mi}\rangle_l^t \, \delta\langle P_{nj}\rangle_l^t}J_{nj}(t), \qquad (7.31)$$

where $J_{nj}(t) = \langle \dot{P}_{nj}\rangle_l^t$ are dissipative flows. To study the evolution of the quasi-temperature of the PA, it is sufficient to highlight the equation with $i = 2$, $m = 1$ in the system (7.31) and save in it two fluxes most significant from a physical point of view. Namely,

$$J_{12}(t) = -L_{\dot{H}_2\dot{H}_2}\{\beta_1 - F_{21}(t)\},$$
$$J_{22}(t) = n_V(N_2, N_2)^t K_V. \qquad (7.32)$$

J_{12} specifies the removal of energy from the subsystem {2} due to phonon scattering of the PA and excitation of the electron–hole pairs. J_{22} flux is associated with a decrease in the energy of the subsystem {2} in the capture of the PA by thermal vacancies, whose density is equal to n_V. The dissipative flux of J_{22} also includes the bulk rate constant for the capture of PA

$$K_V = \begin{cases} \left| v(\bar{\varepsilon}) \cdot 2\pi^2 \left(\dfrac{\lambda_B^2 \Gamma(\bar{\varepsilon})}{\tilde{E}_1} \right) \right| & \text{(phonon scattering)}, \\[4ex] \left| v(\bar{\varepsilon}) 4\pi \left(\dfrac{\lambda_B}{k_F} \right) \right| & \text{(electron scattering)}, \end{cases}$$

where λ_B is thermal de Broglie length; $\Gamma(\bar{\varepsilon})$ is the attenuation of the state of the PA at the Ore band bottom; $\tilde{E}_1$ is the energy of formation of a positron–vacancy complex.

Calculating by the standard procedure [44] the second derivative of entropy, using (7.31) and (7.32) we get

$$\frac{d}{dt}\beta_2(t) = \left[\beta_2(t)\right]^2 \left\{ A(t)\left[\beta_1 - \beta_2(t)\right] + B(t)\right\}, \qquad (7.33)$$

where

$$A(t) = \frac{1}{c(t)}\left(N_2, N_2\right)' L_{\dot{H}_2 \dot{H}_2},$$

$$B(t) = \frac{1}{c(t)}\left(H_2, N_2\right)' K_V n_V,$$

$$c(t) = \left[\beta_2(t)\right]^2 \left\{ \left(H_2, H_2\right)' - \frac{\left(H_2, N_2\right)'\left(N_2, H_2\right)'}{\left(N_2, N_2\right)'}\right\},$$

$c(t)$ is the non-equilibrium heat conductivity of the PA gas, $\beta_2(t) \equiv F_{21}(t)$.

Without additional simplifications it is not possible to find an analytical solution of equation (7.33). We make the necessary changes. First, we neglect the weak time dependence of $A(t)$, $B(t)$ and $c(t)$, by writing them in the limiting case $1/\beta_2 \rightarrow \infty$. This limit, as mentioned above, corresponds to the transition from the values of non-equilibrium thermodynamics to the values of mechanistic statistical theory, in which the occupation numbers of states are constant [128]. Secondly, we introduce a new feature in (7.33)

$$u(t) = \frac{1}{\beta_2(t)} + at,$$

where $a = \beta_1 A + B$. As a result, the new function will satisfy the non-linear ordinary differential equation with constant coefficients

$$\left\{ u(t) - at\right\}\frac{du(t)}{dt} + p = 0,$$

where $p = -A$. If, moreover, t in the resulting equation is replaced by a new variable $\chi = -pt$, with the corresponding change of the functions $u(t)$ to $y(\chi)$, then it takes the form of the Abel equation of the first kind

$$\left[y(\chi) + \frac{a}{p}\chi\right]\frac{d}{d\chi}y(\chi) = 1. \qquad (7.34)$$

The analytical solution of equation (7.34) does exist, but it can not be given explicitly. Imagine a solution

$$\chi = C \exp\left[\frac{a}{p}y(\chi)\right] - \frac{p}{a}\left[y(\chi) + \frac{p}{a}\right], \qquad (7.35)$$

where C is the numerical factor. Recall that in the case of PAs in a dielectric [117] thermalization is almost entirely due to scattering of the PAs on thermal vibrations of the lattice atoms. Taking this into account and assuming that, in general, the capture of the PAs by vacancies makes quite a small contribution to the process of cooling of the subsystem {2}, the shortcoming of solutions (7.35) can be easily 'fixed'. Namely, we obtain (7.35) in the case of the vanishingly small vacancy concentration $n_V \to 0$ (zero approximation) and then using the zero approximation we find

$$\frac{1}{\beta_2(t)} = \left\{\frac{1}{\beta_2} - \frac{2}{3}\beta_1 L_{\dot{H}_2 \dot{H}_2} t\right\}. \qquad (7.36)$$

The solution (7.36) obtained in a rough approximation is applicable only in the initial stage of thermalization, when the process satisfies the condition $t < \left(\beta_1 \beta_1 L_{\dot{H}_2 \dot{H}_2}\right)^{-1}$.

The thermodynamic theory of thermalization presented here does not confirm the starting point [115, 150]. First, on the basis of kinetic equations (7.31) it was found that the PA quasi-temperature satisfies the equation which is converted into a first-order Abel equation (7.34). Naturally, it has nothing to do with the linear equation of parabolic type used in [150]. Second, the exact solution (7.35) was obtained for our equation. On the basis of (7.35) and in the conditions when the capture of the PAS by the vacancies makes a small contribution to the cooling process, it is shown that the temporal course of quasi-temperature (7.36) is close to a linear decrease. The deviation from linearity, according to (7.35), is described by a slowly varying logarithmic function.

We find an analytical expression for τ_T – thermalization time of the PA in dielectrics. It is enough to calculate the quasi-temperature of the PA at the end of the thermalization process. If in (7.36) we substitute $L^{(2)}_{\dot{H}_2 \dot{H}_2}$ in the form (7.12) and neglect the correction (logarithmic) term, then after replacing t and $\bar{\varepsilon}$ we obtain τ_T and $\bar{\varepsilon}_T$

$$\frac{1}{\beta_2(\tau T)} = \frac{1}{\beta_2} - \frac{2}{3}\bar{\varepsilon}_T \tau_T \frac{1}{\tau_2}.$$

Here $\bar{\varepsilon}_T$ is the average energy of the PA in the case of thermal velocities. Now equating $1/\beta_2$ (τ_T) to the thermostat temperature T_1, we find from the equation thus obtained the required expression

$$\tau_T = \frac{3}{2}\left(\frac{1}{\beta_2} - T_1\right)\frac{\tau_2}{\bar{\varepsilon}_T}. \qquad (7.37)$$

Time τ_T (7.37) does not depend on the temperature of the thermostat, since $1/\beta_2 \sim T_1$ (7.27) and $\varepsilon_T \sim T_1$. Estimate of τ_T (7.37) shows that the thermalization time is a few picoseconds and, consequently, much shorter than the annihilation time of PA [140, 143,144,146].

Considering the PA as a wave packet, it is natural to introduce the average group velocity $\bar{v}$ [146]. At the stage of thermalization, when the system goes through a series of successive quasi-equilibrium states corresponding to the current value of time, we can use the hydrodynamic approximation [23,44]. Namely, one can assume that the average velocity of PA $\overline{v(t)}$ depends on time only through the slowly varying quasi-temperature of the PA $1/F_{21}(t)$. In view of (7.36) we get

$$\bar{v}(t) = \left\{\frac{2}{m^*}\frac{1}{F_{21}(t)}\right\}^{\frac{1}{2}} = \left\{\frac{2}{m^*}\frac{1}{\beta_2} - R_1 t^{\frac{1}{2}}\right\}, \qquad (7.38)$$

where $R_1 = 1.33\left(m^*\right)^{-1}\beta_1 L_{\dot{H}_2\dot{H}_2}$. The decrease of the average velocity of PA (P_s) (7.38) and of the free positrons (e^+) is shown in Fig. 7.3. This figure schematically shows the position and width of the Ore gap, together with the fact that the orthopositronium energy is higher than the parapositronium energy when these atoms reach thermal velocities. (In fact, the level of the ground state of orthopositronium lies above the level of positronium by only $8.4 \cdot 10^{-4}$ eV.) The thermal velocity of the PA in the construction of the curve in Fig. 7.3 was assumed to be the value 30 v_s. Finally, it is essential to specify that the curves are not associated with any specific experimental data.

The lifetime of positronium is limited by the annihilation process. Due to the conservation of charge parity, parapositronium annihilates to form two γ-quanta at the time $\tau_s^0 = 1.25 \cdot 10^{-10}$ s. Orthopositronium annihilates with the formation of three γ-quanta at the time $\tau_{tr}^0 = 1.4 \cdot 10^{-7}$ s.

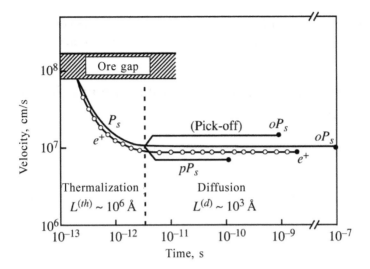

Fig. 7.3. Decrease of the mean velocity of the positronium atoms and free positrons in the thermalization stage. Spatial parameters of free and diffusion motion.

The main mechanism that leads to a marked decrease in the lifetime of orthopositronium is the annihilation on extraneous electrons, called pick-off annihilation. For parapositronium, whose own lifetime is three orders of magnitude shorter than the lifetime of orthopositronium, the frequency of occurrence of the processes, shortening the lifetime, is much lower. Therefore, the lifetime of parapositronium is denoted as τ_s^0, so that the thermalization time $\tau^{(th)}$, the diffusion time $\tau_s^{(d)}$ and τ_s^0 are related by the simple relation

$$\tau_s^0 = \tau^{(th)} + \tau_s^{(d)}.$$

A similar relation can be written for the orthopositronium. However, in this case the physical situation becomes more complicated, since if pick-off annihilation is taken into account the lifetime of positronium in the triplet state can be reduced tens of times. Figure 7.3 shows the main spatial parameters: $L^{(th)}$ – the distance traveled by the PA in the thermalization time, and $L^{(d)}$ – the distance traveled in the diffusion mode. The power temperature dependence of the diffusion coefficient in this case [146] corresponds to the zonal movement of the particles in solids [151,152], whereas an abrupt movement, characteristic of the heavier particles, is described by an exponential function of temperature. The effect of zonal movement was already encountered when discussing the channeling of atomic particles in crystals. However, in this case, we discussed the diffusion of particles in the space of transverse energies.

7.2. Thermodynamics of hot charge carriers in semiconductors

7.2.1. Separation of subsystems into fractions and thermodynamic parameters of fractions

Of course, the results obtained in this section are not universally suitable for all types of semiconductors. Because of the great mathematical difficulties, we restrict ourselves to the study of polar semiconductors under pulsed laser radiation. Our choice is explained by the simplicity of the model. The dependence of transport coefficients on the parameters of the theory can be conveniently studied on the example of GaAs which is a typical polar semiconductor. In gallium arsenide, the minimum of the main conduction band is located at $\mathbf{k} = 0$. Since the valence band maximum is located at the same point, the transitions of charge carriers take place without the change of momentum (the so-called direct transitions), and this greatly simplifies subsequent calculations.

We assume that the system is spatially uniform. Since the free charge carriers in a semiconductor – the conduction electrons and holes – have different masses, we introduce the notation for the mass of an electron and a hole through m_e and m_h. However, we assume that the density of charge carriers is small compared with the density of states in the conduction and valence bands.

Let n_e^0 and n_h^0 be the electron and hole concentrations in the absence of radiation. Under the influence of laser radiation, these concentrations increase by the amount

$$\Delta n_e = n_e - n_e^0, \quad \Delta n_h = n_h - n_h^0.$$

In n-type semiconductors, where $n_e^0 \gg n_h^0$, the change of the number of holes (minority carriers) due to radiation is relatively large, and the number of electrons (majority carriers) increases slightly. In p-type semiconductors, the physical situation is reversed. Note that the process in which there is a large relative increase in the number of minority carriers is known as the injection of minority carriers.

In laser radiation, the semiconductors are characterized by the injection of electron–hole pairs and not of the charge carries of the same type. Moreover, the main injection mechanism during laser irradiation is the generation of pairs through the internal photoelectric effect. The latter is due to absorption of a photon ω_L, exceeding in size the band gap E_G. During injection, the carrier concentration in a certain volume of the crystal increases by Δn_e and Δn_h, which leads to diffusion of excess carriers from this volume.

In addition to the free charge carrier, the system under consideration involves phonons. In this case, a complete description of the kinetics requires the simultaneous solution of the coupled system of kinetic equations for distribution functions of electrons, holes and phonons. Due to great mathematical difficulties, we will not solve this problem in full, and confine ourselves to formulating and solving the relaxation equations.

If the source of radiation is not considered as a separate subsystem, we can assume that the semiconductor consists of two subsystems: the crystal $\{i = 1\}$ and the subsystem of carriers for $\{i = 2\}$. Added to this is that the subsystem $\{i = 2\}$ is naturally divided into two fractions: the electrons and holes. Similarly, the subsystem $\{i = 1\}$ should be split into two fractions: the acoustic and optical phonons. The latter are connected with the special role played by optical phonons in the relaxation kinetics of semiconductors. As a result of the separation of subsystems into fractions, for the shortened description of the semiconductor it is convenient to use the average values of the set of operators P_{mi}

$$P_{11} = \begin{pmatrix} H_a \\ H_{l_0} \end{pmatrix}, \; P_{12} = H_c = \begin{pmatrix} H_e \\ H_h \end{pmatrix}, \; P_{22} = N_c = \begin{pmatrix} N_e \\ N_h \end{pmatrix}. \tag{7.39}$$

Here H_a and H_{l_0} are the Hamiltonians of acoustic and optical phonons

$$H_e = H_e^0 + H_c^{(1)}, \; H_h = H_h^0 + H_c^{(1)}, \; H_c^{(1)} = \begin{pmatrix} H_{ca}^{(1)} \\ H_{c,l_0}^{(1)} \end{pmatrix}, \tag{7.40}$$

where $c = (e, h)$; N_e and N_h are the operators of the electrons and holes; $H_c^{(1)}$ is the interaction Hamiltonian.

To construct non-equilibrium and quasi-equilibrium statistical operators, it is necessary to introduce a set of time non-operator functions $F_{im}(t)$ so that the sum

$$\sum_{im} P_{mi} F_{im}(t)$$

is a dimensionless scalar. Taking into account the form of the operators P_{mi}, the set of non-operator functions can be written as

$$F_{11} = \begin{pmatrix} \beta_1 \\ \beta_{l_0} \end{pmatrix}, \qquad F_{21} = \beta_2, \qquad F_{22} = \begin{pmatrix} -\beta_2 \mu_e \\ -\beta_2 \mu_h \end{pmatrix}, \qquad (7.41)$$

where β_1 – the reciprocal temperature of acoustic phonons; β_{l_0} – the reciprocal quasi-temperature of optical phonons; β_2 – the reciprocal quasi-temperature of free carriers; μ_e and μ_h – the quasi-chemical potential of electrons and holes.

As for the thermodynamic parameters of F_{11}, they can be defined in a different way. In practical applications, the fraction of acoustic phonons is always treated as a thermostat, so the value $1/\beta_1$ can always be taken from experimental data. At the same time, no experimental values are available for the quasi-temperature of the optical phonons. Therefore, in the initial stages of the relaxation process this quantity can be calculated by using the energy balance equation, which will be presented below.

Assuming that the carrier subsystem is in a state of quasi-equilibrium with quasi-temperature $1/\beta_2$, using the traditional procedure [153], we obtain

$$\frac{1}{\tilde{\Omega}} \langle N_c \rangle = n_c^0 \tilde{F}_{1/2}(\beta_2 \mu_c), \qquad n_c^0 = 2 \left(\frac{2\pi m_c}{\beta_2} \right)^{3/2}, \qquad (7.42)$$

where $\tilde{\Omega}$ is volume

$$\tilde{F}_s(\xi) = \int_0^\infty d\zeta \zeta^s \left[\exp(\zeta - \xi) + 1 \right]^{-1}$$

is the Fermi integral.

Let us discuss one more important fact – there are two different scattering mechanisms (7.4) in semiconductors. First, it is scattering of carriers on the acoustic deformation potential. This type of scattering can be considered as elastic, since the change in energy of electrons (or holes) in the elementary act of this scattering is negligible. Second, there is scattering of carriers by optical phonons, the so-called polar optical scattering. The latter is an inelastic process, because the carrier energy is of the order of phonon energy. Inelastic processes provide a more effective channel of energy dissipation.

It is quite reasonable to consider the subsystem of acoustic phonons as a thermostat, as in the elastic scattering heating of this subsystem can be completely ignored. Thus the basic effect of radiation will be in the optical mode of lattice vibrations, which leads to a noticeable heating of the corresponding subsystem, i.e. formation of hot optical phonons.

7.2.2. Kinetic equations for carriers and quasi-temperature of carriers and optical phonons

The time evolution is described by kinetic equations for the thermodynamic parameters (7.41). Given the fact that the temperature of the subsystem of acoustic phonons (the thermostat) is constant, in the same sequence as in [154] we find the following system of equations

$$
\begin{cases}
\left(H_c,H_c\right)'\dot{\beta}_2 - \left(H_c,N_e\right)'\left\{\dot{\beta}_2\mu_e + \beta_2\dot{\mu}_e\right\} - \\
\quad -\left(H_c,N_h\right)'\left\{\dot{\beta}_2\mu_h + \beta_2\dot{\mu}_h\right\} = -J_{12}^{(c)}\left(t\right), \\
\left(H_{l_0},H_{l_0}\right)'\dot{\beta}_{l_0} = -J_{11}^{(l_0)}\left(t\right), \\
\left(N_e,H_c\right)'\dot{\beta}_2 - \left(N_e,N_e\right)'\left\{\dot{\beta}_2\mu_e + \beta_2\dot{\mu}_e\right\} = -J_{22}^{(e,h)}\left(t\right), \\
\left(N_h,H_c\right)'\dot{\beta}_2 - \left(N_h,N_h\right)'\left\{\dot{\beta}_2\mu_h + \beta_2\dot{\mu}_h\right\} = -J_{22}^{(e,h)}\left(t\right).
\end{cases}
\tag{7.43}
$$

The general form of the fluxes $J_{12}^{(c)}$ and $J_{11}^{(l_0)}$, included in (7.43) with (7.39), can be written as follows

$$
J_{12}^{(c)} = J_{12}^{(c,a)} + J_{12}^{(c,l0)},
$$

$$
J_{12}^{(c)}\left(t\right) = \left\langle \dot{P}_{12}\right\rangle_l^t = \mathrm{Sp}\left\{\dot{P}_{12}\,\rho_l\left(t\right)\right\} = -i\,\mathrm{Sp}\left\{\left[P_{12},H_c^{(1)}\right]\rho_l\left(t\right)\right\},
$$

$$
J_{11}^{(l_0)}\left(t\right) = \left\langle \dot{P}_{11}\right\rangle_l^t = \mathrm{Sp}\left\{\dot{P}_{11}\rho_l\left(t\right)\right\} = -i\,\mathrm{Sp}\left\{\left[H_{l_0},H_{c,l_0}^{(1)}\right]\rho_l\left(t\right)\right\}.
\tag{7.44}
$$

In accordance with the formulas of the theory of linear response (section 4.3) we carry out averaging in (7.44) with quasi-equilibrium operator ρ_l. Then the flow of energy in the subsystem of carriers get

$$
J_{12}^{(c)}\left(t\right) = \int_{-\infty}^0 dt'\exp\left(\epsilon t'\right)\left\langle\left[H_c^{(1)}\left(t'\right),\left[H_c,H_c^{(1)}\right]\right]\right\rangle_l^t,
\tag{7.45}
$$

where $\epsilon \to +0$.

The second equation in (7.43) describes the time evolution of the quasi-temperature of the subsystem of optical phonons. In the period in which the energy of charge carriers decreases due to their interaction with optical phonons, the quasi-temperature of the photons increases. As a result, the quasi-temperatures of these subsystems are equalized, i.e. internal thermalization of the carriers and optical phonons takes place. Carriers are cooled by the flux of energy through $J_{12}^{(c,l_0)}$, and optical phonons are heated by the flux $J_{11}^{(l_0)}$, so that we can assume $J_{11}^{(l_0)} = -J_{12}^{(c,l_0)}$. As in the derivation of (7.45), we calculate $J_{12}^{(c,l_0)}$ using the formula of the linear response theory. This gives

$$J_{12}^{(c,l_0)} = -\int_{-\infty}^{0} dt' \exp(\epsilon t') \left\langle \left[H_{c,l_0}^{(1)}(t'), \left[H_{l_0}, H_{c,l_0}^{(1)} \right] \right] \right\rangle^t. \qquad (7.46)$$

Now consider the explicit form of the matrix element of the acoustic deformation potential

$$\left| U_{c,a}(q) \right|^2 = E_c^2 q \left(2n_c \tilde{\Omega} v_s M \right)^{-1}$$

and the matrix element of interaction of carriers with optical phonons

$$\left| U_{c,l_0}(q) \right|^2 = 4\pi \gamma_c^2 \omega_0^2 \left(2m_c \omega_0 \right)^{-\frac{1}{2}} \left(\tilde{\Omega} q^2 \mathcal{E}^2(q) \right)^{-1}.$$

Here E_c is the coupling constant with acoustic phonons; γ_c is the coupling constant with the optical vibrations in the Frohlich model; $\mathcal{E}(q) = 1 + (q_0/q)^2$ is the dielectric constant in the random-phase approximation.

Using the matrix elements mentioned, it is easy to reveal the explicit form of fluxes (7.45) and (7.46). In particular, the flux (7.45), regardless of whether whether the interaction of carriers with acoustic or optical phonons is considered, may be written as

$$J_{12}^{(c,\alpha)}(t) = -2\pi \sum_{c=(e,h)} \sum_{kq} \left| U_{c,a}(q) \right|^2 \left(\varepsilon_{k+q}^c - \varepsilon_k^c \right) \times$$

$$\times \left\{ \left(v^\alpha(q) + 1 \right) f_{k+q}^c \left(1 - f_k^c \right) - v^\alpha(q) f_k^c \left(1 - f_{k+q}^c \right) \right\} \times \qquad (7.47)$$

$$\times \delta \left(\varepsilon_{k+q}^c - \varepsilon_k^c - \omega_q^\alpha \right).$$

Here $c = (e, h)$; $\alpha = (a, l_0)$; f_k^c is the electron distribution function at $c = e$ and holes at $c = h$; $v^\alpha(q)$ is the phonon distribution function; ε_k^c is the energy of charge carriers.

Flux $J_{22}^{(e,h)}$, due to growth of electron–hole pairs in the case of thermal generation, can be written in complete analogy with (7.47) if we introduce the corresponding matrix element $U_{e,h}(\varepsilon_k^c)$ [154]. However, to solve that particular problem, which is discussed below, the flux $J_{22}^{(e,h)}$ is not required in the first approximation. Therefore, its explicit form is not presented here.

We find the final form of the flux (7.47). For this purpose, using the matrix element of the acoustic deformation potential, we run in (7.47) the summation over the variables $\mathbf{k}$ and $\mathbf{q}$. After several transformations, we obtain

$$J_{12}^{(c,a)}(t) = 8\tilde{\Omega}\left(\pi^3 n_c \beta_2^3 \beta_1\right)^{-1}(\beta_2 - \beta_1) \times$$
$$\times \left\{m_e^4 E_e^2 \tilde{F}_1(\beta_2\mu_e) + m_h^4 E_h^2 \tilde{F}_1(\beta_2\mu_h)\right\}, \tag{7.48}$$

where the time dependence of the flux is determined by the time course of the thermodynamic parameters.

As for the flux $J_{12}^{(c,l_0)}$, due to the interaction of carriers with optical vibrations, the exact calculation of (7.46) leads to a cumbersome expression. So it makes sense to introduce the following approximations.

First, we restrict the interaction of electrons with optical phonons. The term of the flux of $J_{12}^{(c,l_0)}$, corresponding to the interaction of holes with optical phonons, is completely omitted. Analysis showed that this term is proportional to the ratio (m_h/m_e), which in the case of polar semiconductors is much smaller than unity.

Secondly, when considering polar optical scattering, screening is not taken into account in all cases. With this in mind, we will consider the contribution of polarization to only one factor, namely, where it significantly influences the flux strength. In this approximation, the flux up to a weakly varying logarithmic term (the last discarded term) can be approximated by the expression

$$J_{12}^{(c,l_0)} = -J_{11}^{(l_0)} = -\frac{1}{\pi^2}K_1\frac{1}{\beta_2}v(l_0) \times$$
$$\times \left[\exp\left(-\beta_{l_0}\omega_0\right) - \exp\left(\beta_2\omega_0\right)\right]\exp\left(\beta_2\mu_e\right), \tag{7.49}$$

where $K_1 = \gamma_e^2 m_e^{3/2}(2\omega_0)^{5/2}\bar{\varepsilon}_e\tilde{\Omega}\left(4\bar{\varepsilon}_e - q_0^2/2m_e\right)^{-1}$, $\bar{\varepsilon}_e$ is the average hot electron energy; $v^{(l_0)} = v(\beta_{l_0}\omega_0)$.

7.2.3. Evolution of quasi-temperature of carriers during thermalization

The initial state, i.e. state from which the process of thermalization begins, is regarded as the quasi-equilibrium in the subsystem of carriers and optical phonons. The latter can be characterized by three main parameters: quasi-temperature of carriers $1/\beta_2$, quasi-temperature of optical phonons and the quasi-equilibrium carrier density $1/\beta_{l_0}^0$ which, according to (7.42), is associated with quasi-chemical potential μ_c.

In the case of laser irradiation of semiconductors, quasi-equilibrium value $1/\beta_2$ is usually determined on the basis of experimental data [155,156]. In particular, in [156] using the ratio of the intensity of luminescence at two different frequencies, the value $(1/\beta_2) = 3000$ K was obtained at a thermostat temperature of 300 K for the GaAs system. In another experiment [155], which also investigated gallium arsenide, it was found that in quasi-equilibrium $(1/\beta_2) = 0.62$ eV $= 7200$ K.

The analytical expression for the hot-carrier quasi-temperature $1/\beta_2$ can be obtained from the energy balance equation for the subsystem {2}. The latter includes the rate of energy pumping by the laser pulse, and loss of energy per unit time as a result of the interaction of carriers with the lattice. It is clear that the losses are determined by the flux $J_{12}^{(c,l_0)}$, which has, as is evident from (7.49), quite a complicated form. So, the resulting transcendental equation can be solved only by numerical integration methods in a specific case. Therefore, we use the experimental value of the quasi-temperature of the hot electrons.

Quite a different case is the definition of the initial quasi-temperature of the optical phonons, since it cannot be taken from the experiment. The theoretical value $(1/\beta_{l_0}^0)$ can be calculated only on the basis of the energy balance equation. As is well known [153] electron–hole pairs are injected into a semiconductor and annihilate in the process of recombination. The binding energy of such a pair is released in the form of phonons or photons. Phonon generation is the dominant process in polar semiconductors as well as in semiconductor compounds and results in pumping of energy into the subsystem of optical phonons. Although the process of warming is not considered here, its result should be considered when calculating $(1/\beta_{l_0}^0)$. Indeed, the increment of the energy of optical phonons due to the recombination of pairs is substituted into the left side of the balance equation and the energy transmitted to the pair by external radiation – to the right side. As a result, we obtain the desired equation

$$\omega_0 \left\{ v\left(\beta_{l_0}, \omega_0\right) - v\left(-\beta^0_{l_0}, \omega_0\right) \right\} = \left\{ \left(\omega_L - E_G\right) - 3\frac{1}{\beta_2} \right\} \langle N_c \rangle.$$

In polar semiconductors under the influence of pulsed laser radiation, the transition of carriers to complete thermodynamic equilibrium occurs in several stages. Of these, two primary stages will be considered in more detail.

In the first stage, as mentioned above, there is a rapid transfer of excess energy of the carriers into the subsystem of optical phonons. This leads to the mutual thermalization of two subsystems: quasi-temperature of the carriers $1/\beta_2(t)$ decreases, and quasi-temperature $1/\beta^0_{l_0}(t)$ grows until it becomes equal to the temperature of the carriers. In the second step further cooling of the subsystem of carriers takes place. The decrease of $1/\beta_2(t)$ is accompanied by a reduction of the quasi-temperature of the optical phonons and the rate of cooling of the subsystem is much smaller than the rate in the first stage.

Using these expressions for the fluxes $J_{12}^{(c,l_0)}$ and $J_{11}^{(l_0)}$, it is easy to write the generalized equation of energy transport [15] in a strongly non-equilibrium system

$$\begin{cases} \left(\dfrac{\overline{dE}\left(\beta_2\left(t\right)\right)}{dt} \right)_{coll} = J_{12}^{c,l_0}\left(t\right), \\[4mm] \left(\dfrac{\overline{dE}_{l_0}\left(\beta_{l_0}\left(t\right)\right)}{dt} \right)_{coll} = J_{11}^{(l_0)}\left(t\right). \end{cases} \tag{7.50}$$

Here $\left(\dfrac{\overline{dE}}{dt} \right)_{coll}$ is the collision term from the generalized equation of energy transfer [154];

$$\overline{E}\left(\beta_2\left(t\right)\right) \equiv \langle H_c \rangle_l^t = \frac{3}{2}\left(\frac{1}{\beta_2\left(t\right)} \right) n_c^0 \tilde{F}_{3/2}\left(\beta_2\left(t\right)\mu_2\right)\tilde{\Omega}$$

is the average energy of carriers in the subsystem at time t;

$$\overline{E}_{l_0}\left(\beta_{l_0}\left(t\right)\right) \equiv \langle H_{l_0} \rangle_l^t = \omega_0 v\left(\beta_{l_0}\left(t\right)\omega_0\right)$$

is the average energy of the subsystem of optical phonons at time t.

The first equation (7.50) describes the cooling of the charge carriers due to scattering by optical phonons, the second – the heating of the optical phonons. Accordingly, we can introduce two relaxation times: τ_{c,l_0} – the energy relaxation time of carriers due to interaction with the l_0-phonons, $\tau_{l_0,c}$ – the energy relaxation time of optical phonons in the interaction with the media. It is interesting to note that the two specified subsystems relax with different times, so that $\tau_{c,l_0} \neq \tau_{l_0,c}$. Indeed, in a particular experiment [154] in the pulsed irradiation of gallium arsenide we have:

$$\tau_{c,l_0} = 5 \cdot 10^{-12} \text{s} \quad \text{and} \quad \tau_{l_0,c} = 10^{-12} \text{ s}.$$

We write the left part of the transport equations (7.50) in the so-called τ-approximation. The relaxation equations (7.50), corresponding to the initial stage of thermalization, then take the following form

$$\begin{cases} J_{12}^{(c,l_0)}(t) = -\dfrac{1}{\tau_{c,l_0}} \left\{ \overline{E}\left(\beta_2(t)\right) - \overline{E}\left(\beta_{l_0},t\right) \right\}, \\[2ex] J_{11}^{(l_0)}(t) = -\dfrac{1}{\tau_{l_0,c}} \left\{ \overline{E}_{l_0}\left(\beta_{l_0}(t)\right) - \overline{E}_{l_0}\left(\beta_2,t\right) \right\}. \end{cases} \tag{7.51}$$

$\overline{E}\left(\beta_{l_0},t\right)$ is the average energy of the subsystem of the charge carriers at the quasi-temperature of another subsystem, i.e. subsystem of l_0-phonons, and $\overline{E}_{l_0}\left(\beta_2,t\right)$ is the average energy of phonons at the quasi-temperature of the subsystems of the carriers.

Equations (7.51) adequately reproduce the time evolution of the quasi-temperature of the carriers during thermalization. However, the complexity of the equations and expressions for the fluxes (7.48) and (7.49) does not enable us to obtain an analytical solution to the problem. Only the numerical solution of equations for the values $1/\beta_2(t)$ is possible in this case.

There is another version of the analysis of thermalization of the carriers. Since the charge carriers in semiconductors are chaotic particles, we can use the Abel equation (7.34). Namely, by applying its analytical solution in the first non-vanishing approximation (7.36), the temporal course of the quasi-temperature of the hot electrons in a polar semiconductor, with (7.49) taken into account, can be written as

$$\left(\frac{1}{\beta_2(t)}\right)_e = \left\{ \frac{1}{\beta_2} - \frac{2}{3}\beta_1 L_{\dot{H}_2 \dot{H}_2}^{(e,l_0)} t \right\}, \tag{7.52}$$

where

$$L_{\dot{H}_2\dot{H}_2}^{(e,l_0)} = \frac{1}{\pi^2} K_1 v^{(l_0)} \left[\exp\left(-\beta_{l_0}\omega_0\right) - \exp\left(-\beta_2\omega_0\right) \right] \exp\left(\beta_2\mu_e\right).$$

Figure 7.4 presents the results of the two variants of the electron quasi-temperature: 1 – on the basis of equations (7.51) (curve a), 2 – using the solutions of the Abel equation (7.52) (curve b). The experimental values of the quasi-temperature of the electrons (dots in Fig. 7.4) were obtained in [156] where an n-type GaAs crystal with a density of $n_e = 10^{19}$ cm^{-3} was studied at room temperature. The crystal was irradiated with laser pulses whose duration was ~6 ps (the second harmonic). The time shown in the figure was measured from the end of the pulse, but the heating of carriers is completely omitted in the graph.

In real experiments the initial stage involved a quick transfer of the excess energy of electrons excited by a laser pulse into a subsystem of optical phonons [154–156]. This process can be described by equation (7.52) provided that the Onsager kinetic coefficient $L_{\dot{H}_2\dot{H}_2}$ includes only the interaction of electrons with l_0-phonons.

The form factor $L_{\dot{H}_2\dot{H}_2}^{(e,l_0)}$ is also written in this form which is proportional to $K_1 \sim \gamma_c^2$. (Remember that γ_c is the coupling constant of the electrons with optical phonons).

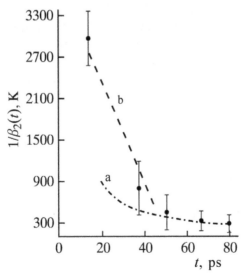

Fig. 7.4. Time dependence of the quasi-temperature of hot electrons in semiconductors in a wide thermalisation temperature range (a) and in the intitial thermalisation stage (b).

As for the acoustic phonons, they only require a negligible fraction of the excess energy and can be ignored at the initial stage. As shown in Fig. 7.4, line b is in satisfactory agreement with the experimental data at short times $t < \left(\beta_1 \beta_1 L_{\dot{H}_2 \dot{H}_2} \right)^{-1}$.

At relatively long times the subcarriers and optical phonons jointly transmit surplus energy to the thermostat, as illustrated by curve a obtained on the basis of equations (7.51). In order to increase the speed of transfer of energy to the optical phonons, it is enough to include in $J_{11}^{(l_0)}$ the contribution of the anharmonicity of third order due to the interaction of a l_0-phonon and two acoustic phonons.

In conclusion, we again return to the question concerning the special role of optical phonons and the thermalization process. Compare the scattering of carriers by acoustic and optical phonons. For scattering on acoustic vibrations we actually deal with the mechanism of elastic scattering, since in this process as a first approximation we can neglect the change in energy of the electron (hole).

However, in the scattering of free carriers on the optical vibrations the phonon energy has the same order of magnitude as the thermal energy of the carrier; the processes in these conditions are inelastic. It is known that the most efficient energy transfer channels in the relaxation kinetics are those which are caused by inelastic scattering processes. From this perspective it becomes clear why the transfer of energy by means of the interaction of carriers with acoustic phonons can be neglected. Indeed, the smaller the energy efficiency of the transmission channel, the slower the relaxation to equilibrium. According to [154], the relaxation time due to interaction with acoustic phonons can be estimated as $(10^{-10}–10^{-9})$ which is far greater than τ_{c,l_0}.

7.3. Displacement cascade and thermodynamics of knocked-on atoms

7.3.1. Cascade function and geometric parameters of the cascade

During a collision the bombarding particles transfer part of their energy to the lattice atoms. If this energy exceeds the threshold displacement energy ε_d, a lattice atom is knocked from the lattice site, i.e. knocked-on atom (KOA). Subsequent collisions of the primary knocked-on atoms (PKOA) with the lattice atoms can lead to the formation of secondary, tertiary KOA, resulting in a cascade of displaced atoms moving in solids. The cascade ends at a time when all KOAs slow down so that their energy drops below the threshold ε_d.

Summarizing the above, a schematic representation of a displacement cascade in the solid is shown in Fig. 7.5. This figure shows the projection on the plane of the trajectory of the primary KOA, trajectories of secondary KOA (the second generation of the displaced atoms), and the trajectories of the third, fourth and fifth generations. Point q_1 is the beginning of the trajectory of KOA, and q_2 is the point of its arrest. In addition, the figure shows schematically the elementary act of scattering in which one of the colliding atoms stops at a node, while the other continues to move in the lattice. Areas of elastic scattering on the trajectory of the KOA are also shown, with scattering leading to the formation of a vacancy and also an additional displaced atom. The process of annihilation of the vacancy and the interstitial atom is also described.

Individual elements of the cascade region are investigated theoretically and experimentally. Models were constructed and field-ion microscopy studies of depleted and enriched zones [157,158] and thermal spikes [159] were carried out. The conditions of formation

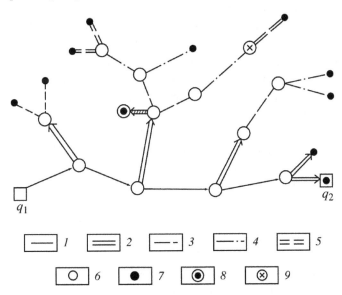

Fig. 7.5. Diagram of the displacement cascade. 1–5) trajectories: 1) primary knocked-on atoms, 2, 3, 4, 5) the knocked-on atoms of the second, third, fourth and fifth generation, respectively; 6) vacancy; 7) interstitial atom; 8) the area of annihilation of the vacancy; 9) the remaining atom.

of the particles moving in the channeling regime [159] were investigated, the conditions of propagation of focusons and dynamic crowdions [132] were studied, the average size of the cascade region was determined [160,161], and the impact of impurities on the cascade development was researched [162].

The energy of the bombarding particles is spent not only on the formation of point defects; much of it is spent on ionization and excitation of the electronic subsystem and, above all on increasing the thermal energy of the conduction electrons. Strong excitation of the electronic subsystem can result in significant heating of the local region inside the solid. Since the behavior of point defects depends on temperature and applied stresses in the region of local heating, they quickly begin to migrate, leaving to sinks (dislocations, grain boundaries, secondary phase separation). In addition, point defects can be combined into clusters, forming thus pores, as well as vacancy and interstitial dislocation loops.

Some of the recombination processes have already been studied. In particular, the volume and shape of the zone of spontaneous recombination of Frenkel pairs were investigated. After irradiation, solids produce a large number of unstable Frenkel pairs, including vacancies and interstitial atoms at distances less than the linear size of the zone of spontaneous recombination. Their concentration is usually more than two orders of magnitude higher than the concentration of stable pairs. Only after the displacement process is fully completed, we can determine which of the Frenkel defects remain stable. Those defects that remain after the completion of the relaxation processes are responsible for changes in the physical properties of the target.

So, these considerations show that the effect of the bombarding particles on solids leads to the formation of KOA that can move through the lattice over many interatomic distances. The result of their collisions with the lattice atoms is the formation of Frenkel pairs, sputtering of the surface and radiation mixing of the components that make up the solid. All these processes occur at velocities of moving atoms many times greater than thermal velocity. The parameters of PKOA can provide information about the mechanism of interaction of high-energy particles with a solid since the inhibition of KOA is the result of a sequence of atomic collisions.

To solve the problem of calculating ν – the average number of KOA attributable to each PKOA – the following model representations are introduced in the classical theory:

– We assume that all collisions are binary, i.e. each moving atom interacts at the given moment with a single atom of the target. This assumption is reasonable, since the interaction force is applied to the area much smaller than the distance between the atoms.

– We assume that all collisions occur only between the moving atom and one of the lattice atoms located in the trest state. Indeed, the density of moving atoms formed in virtually all the available fluxes of incident particles is too small to expect a fairly significant number of collisions between the moving atoms.

– We assume that the KOA loses energy only through ionization as long as its kinetic energy is less than the limiting ionization energy E_i.

– All the KOA with kinetic energy E_c, which is lower than E_i, lose energy only through elastic collisions with lattice atoms and in these collisions behave as rigid spheres [163,164].

– The number of displaced atoms will only increase if both atoms after the collision will have a kinetic energy exceeding ε_d.

Figue 7.6 shows a diagram on which the abscissa gives the energy of the bombarding atom after a collision E_1, and the vertical axis – the energy of the lattice atom after a collision E_2. In this diagram, there are four main areas: A – when both atoms are stopped after the collision, B – corresponds to the act of scattering in which the bombarding atom stops at the site, while the lattice atom departs from the occupied site (there is no additional displacement), C – the lattice atom is in the node, the incident atom continues to move; D – both atoms are emitted from the site (there is an additional shift).

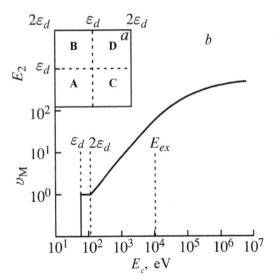

Fig. 7.6. Energy diagram for the case of inelastic collision of the bombarding atom with the atom of the lattice (a). The modified cascade function (b).

In the present model, the average number of KOA which are formed as a result of elastic collisions of the PKOA with the lattice atoms is [122]

$$v(\varepsilon_0) = \begin{cases} 1 & \text{at } 2\varepsilon_d > \varepsilon_0 > 0, \\ \dfrac{\varepsilon_0}{2\varepsilon_d} & \text{at } \varepsilon_0 > 2\varepsilon_d, \end{cases} \tag{7.53}$$

where ε_0 is the energy of the PKOA. According to (7.53), only half the energy of the PKOA is used in displacement of the lattice atoms, the rest is used in collisions with the energy below the threshold.

Further, it is appropriate to consider another definition – the modified cascade function. The fact is that the actual atom–atom collisions do not fully correspond to the model of rigid spheres. Therefore, the KOA forms a secondary KOA with the mean energy, which is considerably lower than half the maximum recoil energy maxε_1, so that the findings are not completely confirmed by the model of rigid spheres. If we make the traditional assumption that the secondary KOA reach such energies at which the excitation of electrons does not occur, the energy used in inelastic collisions is completely determined by the energy losses of the PKOA.

We define the average energy of scattering of PKOA. We have

$$E_{av} = \int_0^{\max \varepsilon_1} dE \left(-\frac{dE}{dx} \right)_c \left[\left(-\frac{dE}{dx} \right)_{ex} + \left(-\frac{dE}{dx} \right)_c \right]^{-1}, \tag{7.54}$$

where $\left(-\dfrac{dE}{dx} \right)_c$ and $\left(-\dfrac{dE}{dx} \right)_{ex}$ are the energy losses of KOA in elastic collisions (7.54) and losses due to excitation of electrons [164]. Using E_{av}, a modified cascade function can be written as [164]

$$v_M = \begin{cases} 1 & \text{at } 2\varepsilon_d > E_{av} > 0, \\ \dfrac{E_{av}}{2\varepsilon_d} & \text{at } E_{ex} > E_{av} > 2\varepsilon_d. \end{cases} \tag{7.55}$$

As an example of using (7.55) we consider the number of displaced atoms formed in graphite. Numerical integration of equations (7.54) and (7.55) gives the result graphically represented in Fig. 7.6b. The figure shows that at energies above the excitation energy E_{ex} the energy losses contribute to electronic excitation, which leads to a deviation from linearity of the curve. Moreover, at energies exceeding $10E_{ex}$, the energy dissipated in elastic collisions grows very slowly with increasing E_c. Unique 'saturation' of the cascade function take place. The saturation effect is observed experimentally (see [165], Fig. 7).

The spatial distribution of defects produced by radiation significantly affects the change in the physical properties of solids. The most interesting qualitative result of the analysis of the spatial distribution of the detection is the difference between the distribution of vacancies and interstitials, or more precisely, the formation of a zone enriched with vacancies (the so-called depleted zone), which is surrounded by a region with a surplus of interstitial atoms.

The calculation of the longitudinal, i.e. in the direction of initial movement of PKOA, as well as the transverse dimensions of the depleted zone is described in [161]. In particular, the empirical dependence of the mean radius of the depleted zone $\tilde{r}$ in the direction perpendicular to its maximum size, on $\bar{\varepsilon}$ – reduced energy of PKOA – was determined; it is shown that $\tilde{r} = 0.16(\bar{\varepsilon})^{1/2}$. It was also revealed that the maximum (longitudinal) size of this zone is consistent with the projected path of the PKOA, calculated on the basis of the Moliere approximation of the interatomic Thomas–Fermi potential [164]. Thus, on the basis of experimental data it can be concluded that the approximation of the depleted zone by a cylinder with the height equal to the projected path of PKOA, and radius $\tilde{r}$ in a wide energy range of PKOA is acceptable.

7.3.2. Transition to the thermodynamic description level

Before proceeding to the energy balance in the early stages of cascade formation, we supplement the above model representations with some provisions of the microscopic theory:

– We consider the cascade process in a continuous medium. Of course, this does not allow us to study the effects caused by the crystal lattice structure, such as focusing substitutional collisions, channeling, and correlated substitutions. However, in most situations of physical interest (e.g. cascades produced by neutron irradiation with energy ~1 MeV), the spatial structure of the cascade has only a minor effect [166] on the energy distribution of the KOA. Moreover, it appears that the effect caused by the crystal structure does not actually affect the number of defects in the cascade. Thus, there are strong indications that the cascade process in a continuous medium can be considered an adequate model of the cascade in a crystalline solid, although this model cannot achieve high accuracy.

– In contrast to [167] we do not require that the bombarded atom leaves the potential well before the shift. Therefore, reproduction of the displacement occurs only when the energy of the incident particle exceeds $2\varepsilon_d$, while at $2\varepsilon_d > \varepsilon > 0$ only subliminal collisions without reproduction of the displaced atoms take place.

– Analytical solutions of equations of the cascade theory were derived obtained using the two interatomic potentials: the potential of rigid spheres [122] and the Lindhard exponential potential [69]. But in terms of applications, both potentials are not as effective as more realistic model potentials: the Born–Mayer potential and the Thomas–Fermi potential. Indeed, the use of, for example, the exponential potential leads to a divergence in the collision term of the kinetic equation. Removal of divergences requires additional regularization with a separate description of the distant and close collisions. Therefore, this section uses the Born–Mayer potential, which allows us to completely remove the problem of divergences.

– Refine the model by taking into account the recombination processes [168,169].

– To simulate the energy removal mechanism from the subsystem of KOA, we consider two possible channels with high rates of energy dissipation. First, the energy removal mechanism due to creation of phonons in the scattering of VA, including the birth of phonons with the non-equilibrium distribution function [170]. This channel is used to remove the kinetic energy, built up in the cascade area, by the phonon mechanism. Second, the mechanism of energy transfer associated with excitation of electrons [171,172]. This mechanism is very effective in metals where electron density is high. Under the influence of a moving atomic particle, the electron gas is excited along its path, and then (in fact, during thermalization) transfers the surplus energy to the lattice through the mechanism of electron-phonon interaction and via non-radiative recombination of electron-hole pairs.

Turning to the level of the thermodynamic description, we study a cascade process on the time scale of heat transfer. Use of this 'coarsened' scale is useful primarily because it allows us to introduce the theory of thermal motion of atoms in the medium and, consequently, the final temperature of the thermostat. Now, assuming that the KOA is a closed thermodynamic subsystem, the complete system should be divided into two subsystems interacting weakly with each other: $\{i = 1\}$ thermostat, i.e. continuous medium, and $\{i = 2\}$ gas of 'hot' atoms, i.e. KOA gas.

In the reduced description, the system will be characterized by a set of average values of only two variables P_{mi}, where $m = 1, 2$,

$P_{1i} = H_i$ is the Hamiltonian and $P_{2i} = N_i$ – the number of particles of the i-th subsystem. The total Hamiltonian of the KOA can be written as $H_2 = H_0^{(2)} + H_{int}^{(2)} = H_i$, where $H_0^{(2)}$ is the Hamiltonian of the gas of non-interacting particles, $H_{int}^{(2)}$ is the Hamiltonian of interaction of the KOA with the thermostat. Moreover, the interaction is associated with the scattering of KOA, which is accompanied by the creation of either phonons or the generation of one-pair excitations. The interaction between identical particles is not considered, since the density of KOA in the adopted model is considered small.

This section is fairly brief, and we will not repeat here the definitions that have already appeared in previous chapters. In particular, the entropy functional can be written in the same form as in Section 7.2, the expansion of the non-equilibrium statistical functional – in the form (7.2) and (7.3), and the thermodynamic equation – in the form (7.4), (7.5).

The relationship of the subsystem quasi-temperature {2} with the fluctuation of energy of the KOA in the non-equilibrium stationary state can be written in the form (7.7). Namely, [173]

$$\left(\frac{1}{F_{21}} \right)_{ss} \equiv \frac{1}{\beta_2} = \left\{ \frac{1}{C_V} (H_2, H_2) \right\}^{1/2}, \qquad (7.56)$$

where C_V is the heat capacity of the gas of KOA per particle. The definition (7.56) is similar to the relations of macroscopic fluctuation theory based on the fluctuation–dissipation theorem and Onsager' hypothesis of the linear decay of fluctuations. This formulation allows the calculation of the thermodynamic parameter of the subsystem of the KOA through a second-order correlation in terms of inelastic scattering [173–175]. If the non-equilibrium processes are non-linear, then the fluctuations are characterized not only by quadratic correlators but also by correlators of higher orders. In this case, the non-equilibrium fluctuation phenomena are studied by methods of non-linear fluctuation thermodynamics. This is beyond the scope of issues we are discussing.

The energy distribution of the KOA will be determined. The early stages of cascade formation end with the establishment of quasi-equilibrium in the subsystem {2}, which is regarded as a non-equilibrium steady state. As in the previous sections, which considered the motion of fast chaotic particles, the stationary distribution of KOA with respect to energy ε with S- and V-contributions taken into account can be written as

$$\tilde{p}(\varepsilon)d\varepsilon = \tilde{G}(\varepsilon)n^{(2)}(\varepsilon)d\varepsilon. \tag{7.57}$$

Here $n^{(2)}(\varepsilon) = \lambda\exp(F_{21}\varepsilon)$ is the average number of filled states in the KOA in the quasi-equilibrium conditions, and $\tilde{G}(\varepsilon)$ is the density of states of the KOA, λ is absolute activity. Analysis of the function $\tilde{G}(\varepsilon)$ over the entire range of ε can be carried out only by numerical integration. In the analytical version of the theory the function can be approximated. In particular, we use an interpolation formula [173]

$$\tilde{G}(\varepsilon) \equiv \tilde{G}(\chi) = \frac{7}{15}\alpha\left\{\exp\left(-C_1^3\chi^4\right)\right\}\left\{\exp\left(3C_1\chi^2 - 1\right)\right\}, \tag{7.58}$$

where $C_1 = 4/7$; $\chi = \left(\varepsilon/\varepsilon^*\right)^{3/4}$; $\alpha = 2MR^2$, $R-$ the parameter of the Born–Mayer potential $U(r) = U_0\exp\{-(r/R)\}$; $\varepsilon^* = (12)^{1/3}\alpha^{-1}$. The function (7.58) (Fig. 7.7) has a maximum, which is typical for the distribution of the density of states of fast particles (in contrast to the particle density close to equilibrium). Note also that for large ε the value of $\tilde{G}(\varepsilon)$ gives the functional relationship similar to the energy distribution of fast particles in Coulomb collisions [164].

7.3.3. The energy balance equation of knocked-on atoms

To construct the energy balance equation of the KOA in the quasi-equilibrium state (or close to it), it is necessary to calculate the power

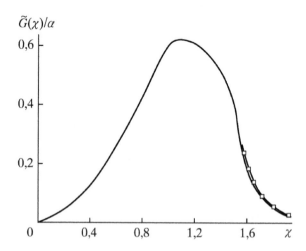

Fig. 7.7. Interpolation function in a wide energy range. The squares indicate the distribution function [164].

supplied by radiation, and the heat dissipated from the subsystem {2}. Removal of energy due to inelastic scattering processes of the KOA is determined by the dissipation function written in the usual form

$$\left\langle \left(\frac{dE_2}{dt} \right)_{out} \right\rangle = \langle \dot{H}_2 \rangle, \quad \dot{H}_2 = \frac{1}{i} [H_2, H_1].$$

Averaging of the flux $\dot{H}_2$ with respect to the distribution (7.57), (7.58) is carried out within the framework of the traditional procedure of the theory of relaxation processes [44]. In the case of weak interaction, when the energy exchange between the subsystem {2} and the thermostat is rather slow, we can restrict ourselves to the relations 'fixed' with respect to the thermodynamic parameters. In this approximation, keeping the second-order term in the interaction potential, we obtain

$$\left\langle \left(\frac{dE_2}{dt} \right)_{out} \right\rangle = (\beta_1 - F_{21}) \int_{-\infty}^{0} dt \left(\dot{H}_2, \dot{H}_2 \right)^t e^{\varepsilon t}. \tag{7.59}$$

Consistent calculation of the average energy input per unit time is based on the theory of the system response to the inclusion of an external perturbation [44]. To do this, we must calculate the addition $\Delta \rho$ to the functional form (7.2) proportional to the $\tilde{\Phi}$-density of the flux of bombarding particles, and then use $\rho + \Delta \rho$ to find the average kinetic energy released (per unit time) in the cascade area. However $\left\langle \left(\frac{dE_2}{dt} \right)_{in} \right\rangle$ can also be calculated by simplified calculations, if from the statistical theory of stages we take the finished form – the cascade function, averaged over the energy spectrum of PKOA. The cascade function that characterizes the total number of ejected atoms per PKOA after averaging becomes [114]

$$\overline{v} = \begin{cases} \dfrac{1}{2} \dfrac{T_m}{T_m - \varepsilon_d} \left(1 + \ln \left(\dfrac{T_m}{2\varepsilon_d} \right) \right) \\ \qquad \text{(bombardment with atomic particles),} \\ T_m (4\varepsilon_d)^{-1} \\ \qquad \text{(bombardment with neutrons).} \end{cases}$$

Here T_m is the maximum energy transferred to the knocked-on atom.

If $\sigma^{(d)}$ is the cross-section of collisions that lead to the displacements [145], the number of PKOA formed resulting in unit time is $\tilde{\Phi}\sigma^{(d)}N_1$. This means that the reproduction rate of the particles is

$$\frac{dN_2}{dt} = \tilde{\Phi}\sigma^{(d)}N_1\bar{v}. \qquad (7.60)$$

Expression (7.60) should be supplemented by a contribution given by spontaneous athermal recombination of Frenkel pairs; the geometry, physical parameters and linear dimensions that characterize the zones of athermal recombination, significantly affect the distribution of defects after the cascade process.

However, these problems lie outside the range of issues addressed at the relaxation stage, and here we can use the simplest model, assuming that the displaced atoms and vacancies recombine whenever come to a distance less than the recombination radius $R_c(\varepsilon)$. We neglect the possibility that more than one displaced atom is found around each vacancy. Then the number of displacements in the damaged volume $\tilde{V}_1$, remaining after recombination, is $\bar{v}\exp\left(-\bar{v}\tilde{V}_2/\tilde{V}_1\right)$ [168,169], where $\tilde{V}_2$ is the volume of spontaneous recombination. If we introduce the recombination cross section $\sigma^{(r)} = \pi R_c^2(\bar{\varepsilon})$, then the full rate of formation of the KOA at $\bar{v}\tilde{V}_2 \ll \tilde{V}_1$ with (7.60) taken into account can be written in the form of a chemical kinetics equation

$$\left(\frac{dN_2}{dt}\right)_{tot} = \tilde{\Phi}\sigma^{(d)}N_1\bar{v}\left\{1-\frac{8}{3}\sigma^{(r)}N_V(\varepsilon_0)a\right\}, \qquad (7.61)$$

where $N_V(\varepsilon_0) = \bar{v}/\tilde{V}_1$ is the density of thermodynamically non-equilibrium formed under the influence of radiation.

The average rate of increase in the internal energy of a subsystem {2} is

$$\left\langle\left(\frac{dE_2}{dt}\right)_{in}\right\rangle = \frac{1}{\overline{N_2}}\langle H_2\rangle\left(\frac{dN_2}{dt}\right). \qquad (7.62)$$

Taking into account the supplied power (7.60) – (7.62) and the rate of removal of energy from the subsystem of the KOA (7.59), the energy balance equation of the KOA can be approximated in the following form

$$\left\langle \frac{dE_2}{dt} \right\rangle = \bar{\varepsilon} f_1 \left(F_{21} \right) \tilde{\Phi} \sigma^{(d)} N_1 \bar{v} \left\{ 1 - \frac{8}{3} \sigma^{(r)} N_V \left(\varepsilon_0 \right) a \right\} -$$
$$- L_{\dot{H}_2 \dot{H}_2} \left(X \right) \cdot X. \tag{7.63}$$

Here

$$\bar{\varepsilon} = \lim_{(1/F_{21}) \to \infty} \left\{ \int_0^{2\varepsilon_d} d\varepsilon \ \tilde{p}(\varepsilon) \varepsilon \right\} = \frac{\pi}{2} \frac{1}{C_1 \alpha} \tag{7.64}$$

is the average energy of KOA in the interpretation of statistical mechanistic theory,

$$L_{\dot{H}_2 \dot{H}_2} \left(X \right) = L_{\dot{H}_2 \dot{H}_2} \left(F_{21} \right) \cdot Y \left(\frac{\bar{\omega}}{2} X \right), \qquad Y(x) = \frac{\text{th} \ x}{x} \tag{7.65}$$

is the transfer coefficient, which depends on the thermodynamic force X [23]; $\bar{\omega}$ is the average energy transfer in collisions at the subthreshold condition in the relaxation phase of the cascade; $L_{\dot{H}_2 \dot{H}_2} \left(F_{21} \right)$ is the transfer coefficient, which is included in a linear relationship between the energy flux and force [23,65].

7.3.4. Quasi-temperature of the subsystem of knocked-on atoms nd thermodynamic stability of states

The quasi-temperature of the subsystem of the KOA is calculated in the same sequence as the quasi-temperature of the positronium atoms (see Section 7.1.3). Using the energy balance equation of the KOA (7.63), we obtain the equation for the fluctuations of the internal energy of the subsystem {2} $\delta E_2(t)$, which is necessary to determine the form of the relaxation equation. (Once again, recall that the fluctuation $\delta E_2(t)$ satisfies the equation of motion, similar to the equation for the mean, but with the additional term $\xi(t)$, which is a Gaussian random variable.)

From the perspective of the general provisions of the fluctuation-dissipation theory [25], the equation for energy fluctuations δE_2 should be supplemented by the equation for the fluctuations in the number of KOA δN_2. However, we can simplify the theory, if we take into account the significant difference of the characteristic times of the problem. The point is that reproduction of KOA occurs at the cascade stage ($\sim 10^{-13}$), whereas the redistribution of energy – at the relaxation stage ($\sim 10^{-12}$ s). In this regard, in the study of irreversible processes on the time scale of heat transfer the details of the kinetics of propagation can be neglected. Therefore, the equation for δN_2

is not used. Note that the same situation occurs in the theory of chemical reactions.

In full accordance with (7.23) the covariance of the random term $\langle \xi(t)\xi(t') \rangle_{AV}$ is expressed through the transfer coefficient in the theory of displacement cascades, namely, the ratio (7.65). This means that the elementary acts that lead to energy dissipation also determine energy fluctuations of the KOA. This, in turn, reflects the basic principle of the fluctuation–dissipation theory: the average values and the fluctuations are determined by the same transitions of the particles.

The value of the quadratic correlator in (7.56) is determined by inelastic scattering. In the first two stages of cascade formation the electrons interact with the fast particle only due to inelastic collisions, which leads to the excitation of the electronic subsystem. Moreover, the electrons remain in the excited state also after passage of the particle. As for elastic collisions, in this case the electrons only screen the nuclear charges (purely adiabatic response of the electronic subsystem) and no excitation takes place. The consequence of excitation of the electrons in the cascade stage is the energy dissipation of the KOA. This is reflected in the notation of the cascade function (7.54), which includes energy loss $(-dE/dx)_{ex}$.

An even greater role is played by the inelastic collisions with the electrons in the relaxation phase of the cascade, when the stationary value of the quasi-temperature of the KOA forms. Indeed, at times $\sim 10^{-12}$ s only scattering processes that are not associated with the transfer of energy to the lattice can take place [164]. Therefore, at this stage, of the two previously mentioned channels of removal of energy (section 7.3.2) only one is effective – the channel associated with the excitation of electrons. The emergence and spread of excitation of electrons manifests itself as a form of heat propagation, and the solid is heated to high temperatures along the path of the KOA. The heated region is expanding rapidly, which, naturally, is accompanied by a decrease in its temperature. This process is known in the literature as the 'thermal spike' [114,150,164]. (It is worth noting that at a later stage of the displacement cascade, namely at the stage of thermalization, the removal of energy from the cascade zone is mainly carried out by the phonon mechanism.)

Restricting ourselves to electron scattering, we introduce the corresponding transport coefficient $L^{(e)}_{\dot{H}_2 \dot{H}_2}(X)$. Now, after selecting the transfer coefficient, it is easy to write the equation for δE_2; this equation is regarded as a relaxation equation and it can be used to obtain the explicit form of matrix relaxation and its eigenvalue. We have

$$h(X) = \frac{1}{C_V} F_{21}^2 \left\{ \overset{*}{\varepsilon}\varepsilon\left(\frac{dN_2}{dt}\right) - L_{\dot{H}_2\dot{H}_2}^{(e)}(X) \right\},$$
(7.66)

where $\overset{*}{\varepsilon} = (2,3)/\alpha$.

Then, within the same approach as that used in Section 7.1.3, we obtain an expression for the variance of the internal energy in the non-equilibrium steady state σ_2^{ss}. Finally, using the definition (7.56) and variance σ_2^{ss}, we obtain the desired equation for the quasi-temperature of the KOA [123]. We have

$$\left(\gamma^2 - 1\right) F_4\left(\gamma; \overline{\omega}\beta_1/2\right) + a_1\left(\beta_1\right) = 0.$$
(7.67)

Here the following notation are used:

$$\gamma = \frac{\beta_2}{\beta_1},$$

$$F_4\left(\gamma; \overline{\omega}\,\beta_1/2\right) = (1-\gamma)\left\{1 - \left(\frac{\overline{\omega}\beta_1}{2}\right)^2 \gamma\right\}^{-1},$$

$$a_1\left(\beta_1\right) = \overset{*}{\varepsilon}\overline{\varepsilon}\frac{1}{\tilde{L}_{\dot{H}_2\dot{H}_2}^{(e)}}\left[Y\left(\frac{\overline{\omega}\beta_1}{2}\right)\right]^{-1}\cdot\frac{dN_2}{dt},$$

$$\tilde{L}_{\dot{H}_2\dot{H}_2}^{(e)} = \lim_{1/F_{21}\to\infty} L_{\dot{H}_2\dot{H}_2}^{(e)}\left(F_{21}\right).$$

The method similar to that used in Section 7.1.3, can be used to calculate the transfer coefficient $\tilde{L}_{\dot{H}_2\dot{H}_2}^{(e)}$ in the relaxation phase of the displacement cascade. To do this, we need first to express the transport coefficient derived by Green's function $G(\mathbf{q}, \omega)$ in accordance with (7.18). Here $G(\mathbf{q}, \omega)$ is the Fourier component of the full Green density–density function (i.e. electron density). This Green function is calculated in the Hartree approximation [149], using the usual relation

$$\frac{1}{G(\mathbf{q},\omega)} = \frac{1}{G^{(0)}(\mathbf{q},\omega)} - U_{eff}(q),$$

where $G^{(0)}(\mathbf{q}, \omega)$ is the Green density–density function of the zeroth order with respect to interaction; $\mathbf{q}$ is the momentum transferred in an inelastic collision of an electron with a KOA, and at the relaxation stage $q \leq 2k_F$; $U_{eff}(q)$ is the effective potential in Hubbard's form [147].

Most of the results concerning the kinetic phenomena in metals, were derived from the isotropic model with a simple conduction band: the electron energy extremum at $\mathbf{k} = 0$ and the spin degeneracy. We will use this model by considering the Fermi gas of high density at zero temperature, $T_1 = 0$. As a result of these transformations, we find the final form of the transport coefficient [123]

$$\tilde{L}^{(e)}_{\dot{H}_2\dot{H}_2} = C_2 \left(Z_1 e^2 \right)^2 \overline{\varepsilon}\, \varepsilon_F \frac{m^2}{M} I_2(\xi),   \tag{7.68}$$

where

$$I_2(\xi) = \ln\left(\frac{v_F}{e^2}\right) - I_3(\xi), \quad I_3(\xi) = \ln\left[\frac{1+3d}{3\pi(1+\xi d)}\right] + \frac{\xi(3-\xi)}{1+\xi d},$$

$$\xi = \left(\frac{k_{TF}}{2k_F}\right)^2,$$

$$d = (2X+1)(4X+1)^{-1}, \quad X = \left(\frac{\omega_p}{4\varepsilon_F}\right)^2, \quad C_2 = \left(10\pi^2\right)^{-1}.$$

Equation (7.67) includes the product of two functions: a slowly varying function $\mathcal{F}_4\left(\gamma; \frac{\overline{\omega}}{2}\beta_1\right)$ and a rapidly changing one (γ^2-1). We use a method that is commonly used in similar cases, namely, we 'fix' the slowly varying function. To this end, we draw attention to the parametric dependence of $\mathcal{F}_4$ on $\frac{\overline{\omega}}{2}\beta_1$. The fact is that in most real situations solids are irradiated at room (or close to room) temperatures. Given, moreover, the characteristic values of $\overline{\omega}$ in the relaxation phase of the cascade, we see that parameter $\frac{\overline{\omega}}{2}\beta_1$ varies in a narrow range close to unity. If for simplicity we fix $\frac{\overline{\omega}}{2}\beta_1 = 1$, then $\mathcal{F}_4 = 1$. Introducing this value of $\mathcal{F}_4$ into (7.67), we find the analytic alsolution

$$\frac{1}{\beta_2} = \frac{T_1}{\sqrt{1 - K\varepsilon \frac{\overset{*}{M}}{m}\left(\frac{dN_2}{dt}\right)}\left[Y\left(\frac{\overline{\omega}\beta_1}{2}\right)\right]^{-1}}.   \tag{7.69}$$

The value of K, included in (7.69) with (7.68) taken into account, takes the simpler form

$$K = \frac{ma_B^2}{C_2 Z_1^2 \varepsilon_F I_2(\xi)}.$$

The coefficient K is applicable in the case of metals with high electron density, and the type of K for non-degenerate n-type semiconductors is given in [123].

Eigenvalue (7.66) is necessary not only for deriving quasi-temperature $1/\beta_2$ (7.69) but also for analyzing the thermodynamic stability of the stationary states of KOA [123,176–178]. Since in this case there is a single eigenvalue, the stability condition takes a rather simple form $h(X) < 0$, which, in turn, gives $a_1(\beta_1) < 1$. To analyze the condition $a_1 < 1$, it suffices to solve the equation (7.63) with $\left(\overline{dE_2} / dt \right) = 0$ with respect to $\tilde{\Phi}$. Then the value of $\tilde{\Phi} / B_2$, where

$$B_2 = 2L_{\dot{H}_2 \dot{H}_2}^{(e)} (F_{21}) \left\{ \langle \varepsilon \rangle \overline{\omega} N_1 \sigma^{(d)} \overline{v} \left[1 - \frac{8}{3} \sigma^{(r)} N_V a \right] \right\}^{-1},$$

can be written as a function of quasi-temperature $1/F_{21}$.

The graph of the function $\tilde{\Phi} / B_2$ at different thermostat temperatures T_1 is shown in Fig. 7.8. The theoretical curves were obtained at $\overline{\omega} = 10^{-1}$ eV, which corresponds to the estimate of the characteristic energy at the relaxation stage. (Although the curves

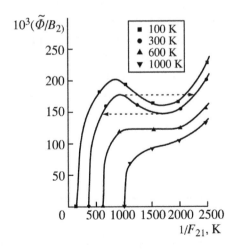

Fig. 7.8. Thermodynamic stability of the stationary states of the knocked-on atoms at the thermostat temperatures of 100, 300, 600 and 1000 K.

were plotted for $\bar{\varepsilon} = 1$ eV, the picture remains qualitatively the same at other $\bar{\varepsilon}$). Depending on what temperature T_1 is, stationary states of two types can be realized in the system, as shown: with one or with three quasi-temperature values.

Isotherm $T_1 = 600$ K is a boundary isotherm separating the isotherms corresponding to a single stationary state and the isotherms with the area of multiple steady states. If $T_1 > 600$ K and the thermostat temperature is kept constant, then for each value of fluence $\tilde{\Phi}$ there is a unique non-equilibrium steady state of the subsystem of the KOA. In this state, the rate of arrival of energy from radiation is balanced by the losses associated with heat transfer. In contrast, when $T_1 < 600$ K, states which lie between the minimum and maximum curve $\tilde{\Phi}/B_2$ are unstable. Naturally, in terms of instability hysteresis phenomena may also be detected, in particular, the arrows at isotherm $T_1 = 300$ K show the hysteresis observed in the transition to the area of multiple stationary states.

If the rate of transfer of energy from the subsystem to the lattice of the KOA is quite low, the internal energy of the KOA increases until the rate of arrival of energy from radiation is not balanced by losses due to heat transfer. Intense pumping of energy can result in high values of the internal energy of the subsystem of the KOA, and this in turn can lead to quasi-temperatures of the KOA well above the lattice temperature. As an example of this situation, we consider the quasi-temperature of the KOA formed in irradiation of light metals. Assume that the exposure occurs at room temperature $T_1 = 1/\beta_1 = 300$ K and the thermodynamic stability condition is satisfied. In this case, using (7.69), we obtain numerical values of the quasi-temperature, which are presented in Table 7.1.

It should be emphasized that the values $1/\beta_2$, presented in the Table, are lower than the melting point of metals T_{mel}, so that the condition $(1/\beta_2) < T_{mel}$ is satisfied. This means that the approach based on the relationships of non-equilibrium statistical thermodynamics indicates the formation is only the heated solid in the central zone of the cascade.

Table 7.1

Element	Z_1	N_1, cm^{-3}	$\dfrac{dN_2}{dt}$, s^{-1}	k_F, cm^{-1}	ω_p, eV	$\dfrac{1}{\beta_2}$, K	T_{mel}, K
Magnesium	12	$4.3 \cdot 10^{22}$	$4 \cdot 10^{14}$	$1.36 \cdot 10^8$	11	950	1520
Aluminiuim	13	$6.0 \cdot 10^{22}$	$6 \cdot 10^{14}$	$1.75 \cdot 10^8$	16	650	933

7.3.5. Discussion of the concept of the quasi-temperature of knocked-on atoms

The temperature of the cascade will be discussed in detail and compared with the melting point of solids. The result of the non-equilibrium thermodynamic theory is described by (7.69) and condition $(1/\beta_2) < T_{mel}$.

In a number of papers dealing with the problem of cascade formation in heavy metals [179,180] T_{ts} – the average temperature of KOA (thermal spike) is determined on the basis of the ratio taken from the classical theory of uniform distribution of energy over the degrees of freedom. In the appendix to the subsystem of the KOA, considered as a monatomic gas (excluding the spatial distribution of energy in the area of the spike [181]), this theorem gives

$$T_{ts} = \frac{2}{3}E_2. \qquad (7.70)$$

Since the expression for internal energy E_2 in practical applications is fairly complicated, further simplifications are introduced into the theory. Namely, in [180] E_2 is replaced by either by the energy or the threshold displacement energy ε_d, or by the energy parameter $(1/2)Mv_s^2$. A more consistent version of the theorem of equipartition of energy and calculation of the internal energy of subsystems of the KOA in irradiation of heavy metals with heavy ions is presented in [182]. Despite some differences in the approaches used in these studies, the main result remains common $T_{ts} > T_{mel}$, so that in the central zone of the cascade there is the liquid phase of the irradiated material (or superheated solid [150]). This finding is consistent with the results of the classical theory of thermal spikes [183] and at the same time with the results of computer simulation [184].

Since the theorem of equipartition of energy has not been proved for the strongly non-equilibrium system, it seems that the relation (7.70) as well as the theoretical conclusions made on the basis of (7.70) should be regarded only as a heuristic attempt to 'guess' the result. Moreover, in the case of pure metals the result $T_{ts} > T_{mel}$ raises serious doubts in terms of the experimentally observed physical situation. The point is that the thermal spike in which there is a region of molten metal should stimulate the process of plastic deformation in this area. However, no actual calculations of this process in specific metals have been carried out, and the experimental data on this issue are very doubtful [114].

Let us assume that the thermal spike can not create a stable molten zone, but it can cause in the comditions $T_{ts} > T_{mel}$ significant local superheating, and although such superheating can be quite intense, it is realized only in a small volume and within a short time. Yet even in these circumstances stimulation of activated processes, in particular, the formation of Frenkel pairs [150], should be quite evident. The kinetics of these processes has not been studied, and most researchers simply assume that the heating in the thermal spikes gives a smaller contribution to the formation of Frenkel defects than collisions in the cascade formation process.

Interesting conclusions concerning the thermal effects, which could be interpreted as a consequence of thermal spikes, were made in [185]. In this paper, attention was given to the atomic ion deposition of coatings, which was accompanied by irradiation with heavy ions and hence by the formation of cascades in the deposited layer. So, no influence of thermal spikes on the structure of the irradiated solid, associated with the spike temperature being higher than the melting temperature, was found in the experiments.

There are two lines of research in solving the problem of thermalization of the cascades in solids: a phenomenological approach [115,150], based on the heat conductivity equation, and the approach proposed in [116], in which the system is described on the thermodynamic level using the concepts of microscopic theory. In this monograph, we have presented only the equations of statistical thermodynamics of displacement cascades.

As already mentioned, the relaxation stage is followed by thermalization of the subsystem of KOA, where the kinetic energy, released in the cascade area, is absorbed by the phonon mechanism. The most intense process at this stage is the scattering of phonons on the KOA and athermal recombination of defects. As for the duration of the final stage, then, given some experimental data (see, for example, [156]), the upper temporal frontier of thermalization t_p of chaotic particles, including the KOA, can be extended to 10^{-10} s.

The decrease of the quasi-temperature of the cascade, calculated on the basis of the Abel equation (7.34), is close to linear, and the deviation from linearity associated with the recombination effect, is described by a slowly varying logarithmic function. To a first approximation, the time course of the quasi-temperature of the cascade is again described by the equation of the form (7.36) and (7.52) if the transfer coefficient in these equations is replaced by (7.12) and (7.13) with the potential of the Born–Mayer atom–atom interaction. As shown by comparison, the pattern of the decrease of the quasi-temperature of the cascade, is in qualitative agreement with the temperature dependence of time, obtained in [184] on the basis of the molecular dynamics method.

The thermodynamic approach to the study of non-equilibrium phenomena in solids seems to us more efficient and better justified. In particular, it can provide justification for the approximations used in more formal statements of the problem, for example, in its phenomenological interpretation.

7.3.6. Kinetics of thermal sputtering

As an example of a linear decrease of quasi-temperature in the form (7.36) we consider the thermal sputtering process – evaporation of lattice atoms, located in the surface layer of solids. In Einstein's model, the lattice atom is a harmonic oscillator oscillating at natural frequency Ω in the potential well of height E_b. It is assumed that $E_b \gg T_2^0$ where T_2^0 is the quasi-temperature of the KOA at the initial time of thermalization. The elementary act of evaporation is then the transition of an atom from the ground state level in the well to the level of the absorbing barrier, and the energy difference between these levels E_b can be interpreted as the binding energy with the surface. The time dependence of the probability of thermal transition is completely determined by the time course of quasi-temperature in the thermal spike zone

$$P\big(\beta_2(t)\big) = \frac{1}{2\pi^{3/2}}\Omega\big(E_b\beta_2(t)\big)^{-1/2}\exp\big\{-\big(E_b\beta_2(t)\big)\big\}.$$

The thermal sputtering coefficient S_{therm} characterizes the number of emitted atoms per one incident ion. If the flux of the vaporized atoms through the free surface is written as $j(t) = N_1 r_s P(\beta_2(t))$, then the sputtering coefficient is given by

$$S_{therm} = \Pi(E_0)\int_0^{t_p} dt\, j(t) =$$

$$= \big(2\pi^{3/2}\big)^{-1} N_1 r_s \Pi(E_0)\Omega E_b \frac{1}{\tilde{K}}\Gamma\left(-\frac{3}{2}, \frac{E_b}{T_2^0}\right). \qquad (7.71)$$

In (7.71) we use the notation: E_0 – energy of the incident ion,

$$N_1 = \frac{3}{4\pi}r_s^{-3}, \qquad \Pi(E_0) = \pi r_0^2\big(r_0 / \lambda(E_0)\big)$$

is the average heated area per one incident ion, in the conditions of chaotic incidence of the beam; $\lambda(E_0) = \{\sigma(E_0) N_1\}^{-1}$ is the mean free path of the bombarding ions, $\sigma(E_0)$ is the cross section; $\Gamma(\alpha, x)$ is the incomplete gamma function. The second line in (7.71) was obtained in the limit $t_p \to \infty$, since the value of the integral (first line) is mainly formed at the lower limit. As for the value of $\tilde{K}$, also included in (7.71), it can be calculated by neglecting the small contribution of recombination. Then, using (7.71), as well as keeping the main contribution (two-phonon processes) at high temperatures, we obtain

$$\tilde{K} = \frac{2}{3}\beta_1\bar{\varepsilon}\tilde{\Phi} = C_3\tilde{\Omega}\frac{\bar{\varepsilon}}{M^2} \int dq\, q^6 \Delta_q \left[\frac{1}{\omega_q^3}|V(q)|^2\right]^2,$$

where $\tilde{\Omega}$ is volume.

An additional physical quantity was introduced in the phenomenological theory of activated processes [150]: n_j – the number of realizations of the activated process over the lifetime of the thermal spike. This quantity is related to the amount of heat Q_2, released within a sphere of radius r_s

$$n_j = \frac{1}{\pi^2}\left(\frac{Q_2}{E_b}\right)^{2/3}\tilde{\rho}. \tag{7.72}$$

After expressing the binding energy by the elastic constant $\bar{g}$, $E_b = (1/\pi^2)\bar{g}\, r_s^2$ we write down the value of $\tilde{\rho}$ in our notation

$$\tilde{\rho} = \Omega r_s^2 \left(\tilde{K}/\bar{g}\right)^{-1}.$$

Using (7.72), we obtain the desired relation

$$S_{therm} = C_4 N_1 r_s \Pi(E_0)\{v(\varepsilon_0)\}^{-2/3} f_1\left(E_b/T_2^0\right)n_j. \tag{7.73}$$

It follows from (7.73) that this equation connects the sputtering coefficient with the number of realizations n_j and the function which depends on the initial thermal spike quasi-temperature

$$f_1\left(E_b/T_2^0\right) = \left(T_2^0/E_b\right)^{11/6}\exp\left(-E_b/T_2^0\right).$$

As seen from (7.73), the sputtering coefficient is proportional to $\tilde{\rho}$. The analogue of the value $\tilde{\rho}$ in the phenomenological theory [150] is considered as the ratio of atomic diffusion capacity to the thermal diffusion capacity of the lattice. The theory does not make it possible to draw clear conclusions about the value of $\tilde{\rho}$. However, in a number of important applications of the theory, e.g. radiation disordering of alloys, the most likely value of $\tilde{\rho}$ is 1 [150]. In order to use representations of chemical kinetics, we simplify our task, assuming $\tilde{\rho} = 1$. Then only one interaction remains in the problem, namely, the interaction of the bombarding ion and the lattice atom, and hence only one cross section $\sigma(E_0)$. In this case, the 'quasi-chemical' reaction, in which we are interested, describes the transformation of particles of one type into another particle

$$A_1 + S\{A_2\} \xrightarrow{k(1/\beta_2)} \text{ products (sputtered atoms)}$$

where A_1 is the bombarding ion; $S\{A_2\}$ is the set of lattice atoms in the thermal spike; $k(1/\beta_2)$ is the rate constant for thermal sputtering. Taking into account (7.72) (7.73), we find

$$S_{therm} \sim k\left(T_2^0\right) = \tilde{B}\left(\frac{T_2^0}{E_b}\right)^{5/2} \exp\left\{-\left(\frac{E_b}{T_2^0}\right)\right\}. \qquad (7.74)$$

(Expressions for $\tilde{B}$ in the particular case of scattering according to the laws of rigid spheres are given in [116].) Comparing (7.74) with the modified equation for the reaction with activation energy E_b in the form of the Arrhenius equation

$$k_n\left(T\right) = \tilde{B}\left(\frac{T}{E_b}\right)^n \exp\left\{-\left(\frac{E_b}{T}\right)\right\},$$

it can be seen that the formulas coincide at $n = 5/2$ and $T = T_2^0$.

Arrhenius kinetics of complex systems in the vicinity of equilibrium has been studied quite extensively [101]. However, in the thermal spike conditions when there is a discontinuity of the thermodynamic parameters of the crystal and the subsystem of the KOA, the theoretical understanding at the same high level is missing. Therefore, justification of the modified Arrhenius kinetics with the exponent $n = 5/2$ should be considered as the main result of the application of the theory of thermalization, more precisely, the temperature dependence (7.36), in the process of thermal spraying. In conclusion, we note that another example of the implementation of the modified Arrhenius kinetics may be biomolecular chemical reactions in the gas phase, in which the behavior of the system (such as plasma) is determined by binary collisions [101].

8

Statistical thermodynamics of light atoms moving with thermal velocity

8.1. Coefficient (constant) of the exit velocity of particles from the potential minimum Above-barrier jumps of hydrogen atoms in FCC metals

Definitions of non-equilibrium statistical thermodynamics have a high degree of generality. Therefore, based on the common definition of the coefficient (constant) of velocity in the thermodynamics we can study from the same viewpoint also the diffusion of atomic particles in the space of transverse energies, and the diffusion of atoms in the configuration space.

The first of these processes is implemented in the channeling of high-energy particles in crystals when the system as a whole strongly deviates from thermodynamic equilibrium. The second process takes place in the case of transport in solids in already thermalized particles or in the case of migration of light impurity atoms. In the latter case, the system is close to complete equilibrium and migrating particles move with thermal velocities.

A theoretical study of the migration of light particles in solids, developed in the framework of the theory of irreversible processes, is faced with some difficulties connected with the description of the elementary act of diffusion. Note the two most characteristic of them. In [186], the formalism of the linear response theory was used to introduce the operator of migration in the site representation, i.e. the operator corresponding to a particle jump from one localized state to the nearest neighbor.

The difficulty that arises with this approach is that the transition matrix element, included in the migration operator, can not be defined completely. It is possible only to estimate the magnitude of the upper and lower limits. However, if diffusion is studied using the formalism of another branch of the theory of irreversible processes, namely, the main kinetic equation, then difficulties are also inevitable. As an example, we can mention the study [187], in which it was proposed that the elementary act should not be described by the migration operator, and the probability of exchange of the atom with its neighbor was introduced into the theory. The kinetic equation, obtained in [187], was so complicated that the analytical methods used to solve it failed.

In this section, we propose a different way of studying the migration of light (impurity) particles in a linear chain of atoms. Some model ideas will be formulated. Let the linear chain consists of N-atoms, the coordinates of interstitials are denote by x_i. Assume that the number of light atoms in the interstitials is small and they are distributed inhomogeneously, so that there is a gradient of the 'concentration' of impurity centers. The field of the one-dimensional lattice, which acts on the interstitials, is described by the potential with period d and potential barrier U_0. In this case, we assume that the barrier height is large compared with the average thermal energy $U_0 \gg 1/\beta_1$.

To describe the impurity center, we introduce one-particle states of the same type as the condition of the one-dimensional oscillator. In the harmonic approximation where the height of barriers separating the potential wells is infinitely large, energy level ω_s is the same for all wells. The energy of the oscillator (light particles) in the s-th excited state is $\omega_s = \Omega s$, where Ω is the natural frequency of the oscillator, s is the vibrational quantum number. At a finite height of the barriers $U_0 = \Omega s_0$ each energy level ω_s blurs into a zone with width $\Delta\omega_s$. However, in the problem of diffusion migration it is sufficient to restrict ourselves to the limiting case of narrow bands without describing them in detail.

The multilevel well, $s_0 \gg 1$, of final height of U_0 can be regarded as a model of the potential well of light impurity atoms (in particular, the hydrogen atom), located in the octahedral positions of the three-dimensional face-centered cubic (fcc) lattice. The analysis shows the potential profile of a light atom in an fcc lattice [188], the potential wells form by the Coulomb interaction between ions of the lattice with an impurity center and the appropriate interaction with the conduction electrons. As for the effects of local deformation of the fcc lattice, they contribute little.

Using the relationships of the general theory of discharge of the particles from the potential minimum, let us consider light particle migration in a simplified one-dimensional model of the crystal. For this purpose we introduce the double well potential. The migrating particle is regarded as a truncated oscillator and can be either in one of the states of the left multilevel well (subsystem {2}) or the right one (subsystem {3}). To describe this alternative, it is enough to use the variables of external degrees of freedom, namely the operators of the occupation number for the states of the left and right harmonic wells c^+c, and d^+d, respectively. These operators have eigenvalues equal to 0 and 1.

The general form factor of the exit velocity of the particles from the potential minimum can be written in the form (section 5.3.1)

$$R = \frac{\pi}{4}|J_{23}|^2 \int_{-\infty}^{\infty} d\omega I(-\omega)K(\omega). \tag{8.1}$$

Here $I(\omega)$ is the spectral intensity of the density–density correlation function (i.e. the number density of light moving particles); $K(\omega)$ is the spectral intensity of the correlation function that includes the variables of external degrees of freedom; J_{23} is the matrix element of the transition {2} → { 3}. The escape of particles initially localized in the left-hand well will be examined assuming that the number of these particles, passed by the above-barrier transition into the right well, is small, so that the condition of low density $c c^+ \approx 1$ is satisfied. Then, neglecting the interaction between the particles and lattice atoms, it can easily be shown that the correlation function, including the occupation numbers, i.e. variables of the external degrees of freedom, has the form

$$\left\langle c^+d \exp(iH_0t) d^+c \exp(-iH_0t) \right\rangle^{(0)} = \int_{-\infty}^{\infty} d\omega \exp(-i\omega t)K(\omega). \tag{8.2}$$

Here H_0 is the Hamiltonian of two subsystems {2} and {3}.

We shall use the definition (8.1), since the calculation of spectral intensities was analyzed in detail in section 5.3.1. We calculate $I(\omega)$

and $K(\omega)$ in the one-dimensional model of migration and substitute these expressions in (8.1). As a result, the constant of the exit velocity of the particles from the potential minimum of the subsystem $\{2\}$ in the case of weak interaction $\gamma_s^2 \ll \Omega^2$, taking into account (8.2), becomes [188]

$$R = \frac{1}{8}|J_{23}|^2 \frac{1}{\Omega^2} \sum_{s''} \sum_{s \le s_0} \tilde{B}_{s''} \tilde{\gamma}_{s''} \, {}_1F_1\left(-s'', 1; r'(s)\right). \qquad (8.3)$$

Here

$$\tilde{\gamma}_{s''} = \frac{\omega_{s''}}{\Omega} \frac{1}{\tau_3}, \quad \frac{1}{\tau_3} = 2\pi \sum_q \left|\tilde{I}_q\right|^2 \tilde{\varphi}\left(\beta_1 \omega_q\right) \delta\left(\Omega - \omega_q\right),$$

$$\tilde{\varphi}(x) = \left(\exp x - 1\right)^{-1},$$

where $\tilde{I} = \varkappa_0 \left(NMm\Omega\omega_q\right)^{-1/2} \cos\left(q\dfrac{d}{2}\right)$, $\varkappa_0$ is the coupling constant of the particle with the lattice, m is the mass of light impurities, ω_q is phonon energy. Coefficients $\tilde{B}_{s''}$, included in (8.3), satisfy the equation [188]

$$\sum_{s''} \tilde{B}_{s''} \, {}_1F_1\left(-s'', 1, r'\right) = 2n_s, \qquad (8.4)$$

where n_s is the equilibrium distribution of migrating atoms.

Equation (8.3) will be transformed. First, we define the value of r', occurring in (8.3) and (8.4). In problems of high-temperature diffusion, the values of $\xi = \beta_1 \Omega \ll 1$ are of greatest interest in which $r' \approx s$ [188]. Second, in the semi-classical approximation we replace ω_s and r' by corresponding continuous variables ε and $r' = \beta_1 \varepsilon$. To ensure rapid convergence of the series in s'', in (8.3) we use the procedure that was used in [57]. Namely, we replace s'' by the discrete value $\tilde{\eta}_r$ – the root of this transcendental equation is presented below. The sum with respect to s'' and s in (8.3) then becomes

$$\frac{1}{\tau_3} \sum_r \tilde{B}(\tilde{\eta}_r) \tilde{\eta}_r \int_0^{U_0} d\varepsilon \, \tilde{g}(\varepsilon) \, {}_1F_1\left(-\tilde{\eta}_r, 1; r'(\varepsilon)\right), \qquad (8.5)$$

where $\tilde{g}(\varepsilon)$ is the density of states of migrating particles.

Determining the range of variation of the the free variable r' (ε) in (8.5), we must consider the following. According to [186], in a state close to equilibrium, the distribution of particles in the levels of the well is of the Boltzmann type, 'truncated' at the value of ε, equal to the threshold energy $\overset{*}{\varepsilon} = U_0 + \tilde{\varepsilon}$, where $\tilde{\varepsilon}$ is a small increment. We have

$$\tilde{n}(\varepsilon) = \begin{cases} \exp\{\beta_1(\mu_2 - \varepsilon)\} & \text{for } \varepsilon \leq \overset{*}{\varepsilon}, \\ 0 & \text{for } \varepsilon > \overset{*}{\varepsilon}. \end{cases}$$

The chemical potential of migrating particles μ_2 is determined from the normalization condition of the particles. Since the 'truncation' of function $n(\varepsilon)$ occurs in the region of large values of ε, we assume that the dimensionless energy $\beta_1 \varepsilon$ in the equation for $\tilde{B}(\tilde{\eta}_r)$ is such the condition $r'(\varepsilon) \gg 1$ is satisfied. Then in equation (8.4), where the coefficient $\tilde{B}_{s^r}$ is replaced by function $\tilde{B}(\tilde{\eta}_r)$, the hypergeometric function can be replaced by its asymptotic value $\left\{ {}_1F_1\left(-\tilde{\eta}_r, 1; r'(\varepsilon)\right) \right\}_{asymp}$ with $r'(\varepsilon) \to \infty$. The result is a modified equation for the coefficients

$$2\Lambda_2 \exp\{-r'(\varepsilon)\} = \sum_r \tilde{B}(\tilde{\eta}_r) \left\{ {}_1F_1\left(-\tilde{\eta}_r, 1; r'(\varepsilon)\right) \right\}_{asymp}, \qquad (8.6)$$

where $\Lambda_2 = \exp(\beta_1 \mu_2)$. The transcendental equation for $\tilde{\eta}_r$ is easy to derive, starting from the right side of the form (8.6) and the step distribution function $\tilde{n}(\varepsilon)$. We have

$$\left\{ {}_1F_1\left(-\tilde{\eta}_r, 1; r'(\overset{*}{\varepsilon})\right) \right\}_{asymp} = 0. \qquad (8.7)$$

If the expansion $\left\{ {}_1F_1(-\tilde{\eta}_r, 1; r') \right\}_{asymp}$ we retain the 'older' terms and reject the corrections of the order of $1/\left(\overset{*}{\varepsilon}\beta_1\right)$, then the roots of the transcendental equation can be written as

$$\tilde{\eta}_r(\zeta) = (r-1) + \{(r-1)!\}^{-2} (\zeta)^{2r-1} \exp(-\zeta), \qquad (8.8)$$

where $\zeta = \beta_1 \overset{*}{\varepsilon}$.

Obviously, the first root of (8.8) $\tilde{\eta}_1$ is the smallest. In particular, at $\zeta \sim 10$, the order of magnitude of the smallest root η_1 with respect to the successive roots $\tilde{\eta}_2, \tilde{\eta}_3 \ldots$ does not exceed 10^{-3}–10^{-4}. This fact predetermines the essential feature of the sum in (8.3). The point is that the density–density correlation function is associated with the Green density–density function [44]. The latter has the asymptotic form characteristic of the time functions of the stochastic theory in the limit of large times t, for which the method for calculating the kinetic coefficients by comparing the kinetic equation is justified. With such a large t and if $\left(\tilde{\eta}_1 / \tilde{\eta}_{r''}\right) \le 10^{-3} - 10^{-4}$ (where $r'' = 2, 3, 4,...$) the sum of the series (8.5) of corresponding $\tilde{\eta}_2, \tilde{\eta}_3 ...$ is negligible compared with the first term. Retaining in (8.3) the first term, we conclude that for the calculation of R it is sufficient to find the first coefficient $\tilde{B}(\tilde{\eta}_1)$ (not a full set of $\tilde{B}(\tilde{\eta}_r)$). Thus, we find that to calculate (8.3) we must calculate the following expression

$$\tilde{B}(\tilde{\eta}_1) \int d\varepsilon \, \tilde{g}(\varepsilon)\left\{ {}_1F_1\left(-\tilde{\eta}_1, 1; r'(\varepsilon)\right)\right\}. \tag{8.9}$$

Equation (8.7) can be solved by numerical integration, but in two areas of $\zeta = \beta_1 \Omega$ the solution can be written in the analytical form

$$\tilde{\eta}_1(\zeta) = \begin{cases} \zeta\exp(-\zeta) & \text{for } \zeta \gg 1, \\ \zeta\exp(-\zeta)\left[1 + 2\,\mathrm{Ker}(\zeta/2, 2)\right] & \text{for } 7 > \zeta > 2, \end{cases} \tag{8.10}$$

$\mathrm{Ker}(x)$ is the Kelvin function.

Thus, taking into account (8.5), (8.6) and (8.9), we obtain the final expression for the coefficient (constant) of discharge velocity of the particles from the potential minimum

$$R = \left|J_{23}\right|^2 \tilde{\eta}_1(\zeta)\tilde{\varphi}(\xi)(2\tau_1)^{-1}\Omega^{-2}, \tag{8.11}$$

where

$$\left(1/\tau_1\right) = \left(\frac{\tilde{\Omega}_0}{2\pi}\right)\varkappa_0^2\left(M\,mv_s^3\right)^{-1}$$

is the energy relaxation frequency, $\tilde{\Omega}_0$ is the volume per one site.

Using (8.11), we study the diffusion of hydrogen atoms (HA) in fcc metals. First, let us recall the structure of the configuration relief of the HA. As an example, imagine the relief of HA in a copper crystal (Fig. 8.1): the space diagonal of the unit cell has a deep potential well in the center of the cube (the octahedral position O) and two small wells in the tetrahedral (T) positions. In the well, located in T-position, there is a local level so that HA can be localized there [189]. Near the top of the deep well (O-position), the energy levels are condensed, so that the transitions between excited states of the HA are quite realistic in terms of one-phonon scattering. This fact is important, since the relaxation frequency $(\tau_1)^{-1}$ is calculated for the case of the one-phonon process.

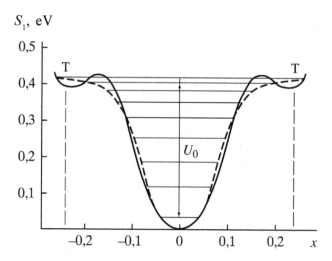

Fig. 8.1. Potential relief of the hydrogen atom along the volume diagonal of the elementary cell of the FCC metal (H–Cu system).

The shape of the potential relief shows that the fcc structure is characterized by above-barrier hopping of the HA along the interstices of different types O–T–O with activation energy U_0 [188]. As for the direct 'jump' from the O-positions in the center of the cube to the O-position in the middle of the cube face, this transition is not realized, since in this case the particle would have had to overcome a much greater barrier than in the O–T–O transition.

$\eta_1(\zeta)$ from (8.10) (when $\zeta \gg 1$) is substituted into (8.11). Then for $\zeta = \beta_1 U_0 \gg 1$ the rate coefficient of above-barrier O–T–O transitions takes the form

$$R^{(f)} = \frac{1}{2}|\tilde{J}_{23}|^2 \zeta \exp(-\zeta)\tilde{\phi}(\xi)(\tau_1 \Omega^2)^{-1}. \tag{8.12}$$

Writing the explicit form of the square of the matrix element of transition $\{2\} \rightarrow \{3\}$, it should be remembered that the HA offsets the nearest neighbor atoms, and, strictly speaking, the rigid lattice model cannot be used. In its simplest form, we have

$$|\tilde{J}_{23}|^2 = \frac{1}{2}\left(\frac{\Omega}{2\pi}\right)^2 \text{Erfc}\left(\frac{U_{eff}}{\sqrt{2}\sigma}\right),$$

where U_{eff} is the difference between the barrier at the saddle point and the elastic energy of the displacement of lattice atoms under the influence of HA; σ^2 is the fluctuation of the square of the barrier height, caused by fluctuations in the phonon subsystem.

8.2. Tunneling transitions of hydrogen atoms due to the collapse of local deformation in the bcc lattice

Full derivation of the equation for the configuration relief of the potential of HA in metal is contained in [190,191]. As shown in these papers, the main contribution to the relief is provided by the Coulomb interaction of metal ions with the impurity center in a uniform 'hard' background of free electrons. If the main amendment which takes into account the contribution of the polarization of the electron gas due to the interaction with the impurity center is added to the main term, the configuration relief takes the form

$$S^{(1)}(R) = \frac{1}{\tilde{\Omega}_0} \sum_{K_n}{}' \frac{4\pi e^2}{K_n^2} \left\{ Z + V(K_n) \frac{\pi(K_n)}{\varepsilon(K_n)} \right\} \cos(K_n R). \qquad (8.13)$$

Here Z is the ion charge; K_n is the reciprocal lattice vector; $V(K_n)$ is the Fourier component of the potential between the electron and the ion; $\pi(K_n)$ and $\varepsilon(K_n)$ is the polarization operator and the dielectric constant, respectively. Site $K_n = 0$ is excluded because of the condition of electroneutrality of the system in general. Expression (8.13) depends on the radius vector R of the proton and the lattice structure, provided that the ions are in ideal lattice sites.

Due to rapid decay of the screened potential of the proton-ion interaction, the sum (8.13) has the weak convergence; in fact, decrease the Fourier components, proportional to K_n^{-2}, is compensated by an increase in the number of nodes, which is proportional to K_n^2. To improve the convergence, equation (8.13) is presented in the nodal form. However, as shown by analysis, $8 \cdot 10^3$ nodes must be taken into account in this case to achieve convergence.

Since the hydrogen in metals is in the quasi-atomic state with a localized electron around a proton, the full potential relief of the HA includes, in addition to (8.13), the contribution of the screened interaction of the localized electron with the ions of the lattice $S^{(2)}(R)$. However, accounting for $S^{(2)}(R)$ leads only to a smoothing of the relief which gives (8.13). Therefore, the amendment made by $S^{(2)}(R)$ is not considered and this approximation using (8.13) is used to compute the topology of the configuration relief on the face of a unit cell of the bcc metal for the Nb – H system (Fig. 8.2).

The origin of the energy in the graph is represented by the bottom of the potential well in the tetrahedral (T) position, and lines of

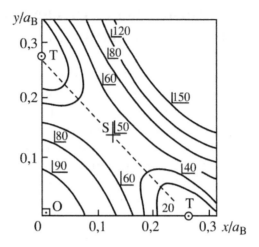

Fig. 8.2. Topology of the potential relief of the hydrogen atom (isoenergetic curves) at the face of the bcc cell (Nb–H system).

equal energy are given in units of 10^{-3} eV. Tetrahedral positions are separated by potential barriers, and the range of the T–T transition has a saddle point S, located in the vicinity of the triangular position of the bcc lattice. Octahedral (O) positions also correspond to saddle points, namely the potential decreases in the O–T direction, but increases in the direction perpendicular to it. This is proof of the low probability of localization of the HA in the O-positions.

The coordinate of the T-position in the upper left corner in Fig. 8.2 is denoted by χ (subsystem {2}), and the coordinate of the other T-position by χ' (subsystem {3}). The potential well at the point χ is denote as $v_\lambda(\chi)$, where the index λ indicates the absence of local deformation. Given the curvature of the bottom of the harmonic well, it is easy to calculate the diagram of its energy levels.

Calculations performed in the harmonic approximation show that in the well $v_\lambda(\chi)$ there is only one level $E_1 = \omega_1$ (in the case of the Nb–H system $E_1 = 2 \cdot 10^{-2}$ eV), and if it is occupied, then the total length of the wave function of the HA coincides with the width of the potential barrier between the wells $v_\lambda(\chi)$ and $v_\lambda(\chi')$. This fact predetermines the high probability of tunneling of the HA between the levels of the ground state of the potential wells $v_\lambda(\chi)$ and $v_\lambda(\chi')$. Nevertheless, the delocalization of the HA during the time $\sim 2\pi/\omega_1$ still happens. The point is that the interaction of the HA with the strain field around an impurity atom forms a deep potential well in which the HA is localized (self-trapping effect). Moreover, as shown by the estimates, the region of localization in this case is smaller than the size of the bottom of the deep potential well.

It follows that HA can perform the tunneling transition $\{2\} \rightarrow \{3\}$, if due to thermal fluctuations the local deformation will be disrupted so that the neighboring ions occupy sites of the perfect bcc lattice. After the transition the local deformation captures the HA in a new T-position. The rate coefficient, given by equation (8.12), is modified so that it matches the tunneling transition of this type. It is sufficient to replace in (8.12) the probability of excitation of the HA $\eta_1(\zeta)$ by the probability of thermal fluctuations $\mathcal{P}(\beta_1)$ disrupting deformation.

Assuming the Gaussian distribution of fluctuations, we have

$$\mathcal{P}(\beta_1) = (2\pi)^{-1/2} e^{-\beta_1 w_{\lambda\lambda'}},$$

where $w_{\lambda\lambda'}$ is the minimum work required to transfer the subsystem $\{2\}$ from the equilibrium state in the presence of local strain (index λ') to the state in the same T-position, but without strain. Naturally, J_{23} in (8.12) should be replaced by the matrix element of the direct tunneling transition $J_{\chi\chi'}$. In the direct transition is no displacement of ions during tunneling. As a result, the general expression (8.12) gives

$$R^{(p)} = (8\pi)^{-1/2} |J_{\chi\chi'}|^2 e^{-\beta_1 w_{\lambda\lambda'}} \frac{1}{\tau_1 \omega_1^2}. \tag{8.14}$$

In this formula, the value $(\tau_1 \omega_1^2)^{-1}$ describes the contribution of dynamic failure of the tunnel state of the HA due to phonon scattering. Since this tunneling occurs between the levels of the ground state (recall that in the wells $v_\lambda(\chi)$ and $v_\lambda(\chi')$ there is only level E_1), then $J_{\chi\chi'}$ in (8.14) can be replaced by a constant value $J_{\lim}$, using the approximation [192–195]

$$J_{\chi\chi'} = J_{\lim} \theta(q_s - q_{sc}), \tag{8.15}$$

where q_{sc} is the critical symmetric vibrational mode q_s.

The work $w_{\lambda\lambda'}$ is computed using the description of the lattice in a 'jelly' model. It is assumed that the ion charge is evenly spread over the volume of the deformed unit cell. The potential of interaction of the HA with the displacement field is given by [191]

$$V_{\lambda'}(R) = -\frac{1}{\tilde{\Omega}_0} Z^* e^2 \int dr \frac{e^{-\alpha|r-R|}}{|r-R|} \text{div } u(r), \tag{8.16}$$

where $u(r)$ is the displacement of an element of the medium at point r; Z^* is the effective charge which in our theory is a free parameter. If the wave function $\varphi(R)$ corresponds to the state of the HA, then the energy of the migrating atom can be represented by the following functional

$$\varepsilon_\lambda[\varphi] = \frac{1}{2m} \int dR \{\nabla\varphi(R)\}^2 + \int dR\, \varphi^*(R) V_\lambda(R) \varphi(R). \qquad (8.17)$$

In addition to (8.17), the total energy includes the energy of elastic deformation

$$W_\lambda[u] = \frac{1}{2}\tilde{K} \int dr \{\text{div } u(r)\}^2, \qquad (8.18)$$

where $\tilde{K}$ is the bulk modulus. The extremum of the functional of total energy $\mathcal{E}_{\lambda'}[\varphi, u]$ with respect to the lattice gives an expression for div $u(r)$. Substituting this expression into (8.16)–(8.18), we represent the total energy functional as a functional of the wave function of the HA. We get

$$\mathcal{E}_\lambda[\varphi] = T[\varphi] + \tilde{E}_{pot}[\varphi] = \frac{1}{2m} \int dR \{\nabla\varphi(R)\}^2 + \tilde{E}_{pot}[\varphi],$$

$$\tilde{E}_{pot}[\varphi] = -\frac{1}{2}\left(\frac{Z^*e^2}{\tilde{\Omega}_0}\right)^2 \frac{1}{\tilde{K}} \int dr\, dR\, dR'\, |\varphi(R)|^2 \times \qquad (8.19)$$

$$\times \frac{\exp\{-\alpha(|R'-r|+|R-r|)\}}{|R'-r||R-r|}|\varphi(R')|^2.$$

The calculation of the total energy is reduced to the functional (8.19) taking into account the normalization conditions for the wave function $\varphi(R)$.

Taking into account that the trial wave function in the ground state should have no nodes, it is convenient to use the hydrogen-like wave function of the 1S-state, namely, $\varphi^{(1)} = \gamma^{3/2}\pi^{-1/2}e^{-\gamma R}$. Then, substituting $\varphi^{(1)}$ into (8.19), after a series of transformations we obtain the energy of the ground state of the self-consistent state of the HA:

$$\mathcal{E}_1(H) = \frac{\gamma^2}{2m} - \frac{\pi}{4\tilde{K}\alpha}\left(\frac{Z^*e^2}{\tilde{\Omega}_0}\right)^2 \sigma^3 \left\{\frac{3}{2}\sigma^2 + \sigma + \frac{1}{2}\right\}, \qquad (8.20)$$

$\sigma^{-1} = 1 + (\alpha/2\gamma)$, $\alpha = a_B^{-1}$. Variational parameter γ, included in the wave function, determines the degree of localization of the HA, due to

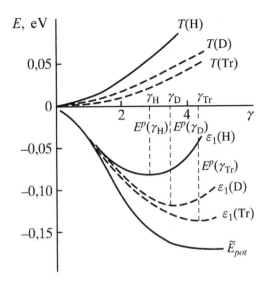

Fig. 8.3. Dependence of the energy of the ground state of the hydrogen atom in the bcc lattice on localisation parameter γ. (Parameter γ is measured in the inverse Bohr radii).

lattice deformation. According to (8.20) (Fig. 8.3), in the case of non-localized states, when $\gamma \to 0$, the potential energy of the system tends to zero, which corresponds to an ideal lattice. At the same time in localization at point $\gamma \to \infty$ the potential energy tends to a constant value $(3\pi/4)(1/\tilde{K}\alpha)(Z^*e^2/\tilde{\Omega}_0)^2$, which is clearly seen in Fig. 8.3. The minimum value of the energy of HA (8.20) forms as a result of competition: the kinetic energy T (H) increases and the potential $\tilde{E}_{pot}$ (H) decreases with increasing localization.

The work $w_{\lambda\lambda'}$ is represented by the energy difference of the system in the ground self-consistent state λ' with local deformation and in the non-localized state λ without deformation. We have

$$w_{\lambda\lambda'} = \min \varepsilon_1 (\mathbf{H}) \equiv E^{(p)}(\gamma_H). \tag{8.21}$$

Given (8.15) and (8.21), the final expression for the tunneling rate (8.14), caused by the collapse of local deformation, has the form

$$R^{(p)} = (8\pi)^{-1/2} |J_{\lim}|^2 \exp\left[-\beta_1 E^{(p)}(\gamma_H)\right] \frac{1}{\tau_1 \omega_1^2}. \tag{8.22}$$

By analogy with the calculation of the parameters of the HA we can also calculate kinetic energies T (D) and T (Tr), total energies ε_1(D) and ε_1(Tr), as well as the minimum work $E^{(p)}(\gamma_D)$ and $E^{(p)}(\gamma_{Tr})$

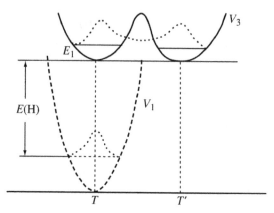

Fig. 8.4. 'Passive' transition of the hydrogen atom in bcc metals.

for deuterium and tritium. The results are shown in the figure, where all parameters are calculated for the systems (Nb–H), (Nb–D) and (Nb–Tr).

HA tunneling due to the collapse of local deformation [190,191] is illustrated in Fig. 8.4. Let it be that the HA occupies the level E_1 in the well V_3, created by the Coulomb forces. It is assumed that the characteristic length of the spatial variation of its wave function coincides with the width of the potential barrier. Due to this, the probability of tunneling (sub-barrier) transition between the levels of the ground state in the wells of T and T' is close to unity. Lattice relaxation in the immediate vicinity of the HA leads to its self-trapping. Indeed, the interaction of HA with the strain field results in the formation of a deep potential well V_1 in the T position. After lattice relaxation, the HA goes to the ground level well V_1 and in this state the localization region of the HA is smaller than the bottom of the potential well.

Based on the analysis of states of the HA, it is natural to assume that hydrogen can carry out transitions at those moments when, due to thermal fluctuations, the ions of the lattice adjacent to the HA occupy sites close to the sites of the ideal rigid lattice. After tunneling T → T' local deformation affects the impurity atom in the new T'-position. The activation energy of the lattice $E^{(p)}$ in such a transition is represented by the difference between the energy of level E_1 in the well of the rigid lattice V_3 and the ground state level in the well V_1, produced by deformation effects

$$E^{(p)} = E_1 + E(H).$$

8.3. Thermally activated tunneling in bcc metals

Let us consider in more detail the energy levels of the deep well $v_\lambda(\chi)$ formed as a result of self-trapping of the HA. The analysis of the harmonic approximation showed that in the 'frozen' well $v_\lambda(\chi)$ there are two levels: ε_1 – the level of the ground state, self-consistent with the well potential $v_\lambda(\chi)$ (HA wave function $\psi^{(1)}$), and $\tilde{\varepsilon}_2$ – level of the excited state which is not self-consistent with the well potential $v_\lambda(\chi)$ (the wave function $\psi^{(2)}$). Transitions can form between these levels of the 'frozen' well. In particular, such transitions can form in the conditions of quasi-elastic scattering of slow neutrons on the HA.

For example, in the niobium–hydrogen system (Nb–H), the first excited level, observed by neutron scattering [196], corresponds to the energy range of $\tilde{\varepsilon}_2 - \varepsilon_1 = 0.114$ eV, which agrees well with the theoretical estimate of 0.11 eV which we obtained in the harmonic approximation. Naturally, the restructuring of local deformation and the formation of a self-consistent excited state whose energy is lower than $\tilde{\varepsilon}_2$ occur after thermal stimulation.

The presence of the excited state of the HA in the subsystem {2} determines the feasibility of another mechanism of elementary diffusion of the HA in bcc metals. Suppose that, initially, the HA is in the ground state. The first phase of the jump is a transition from the ground state $\psi^{(1)}(\chi)$ to the excited state $\psi^{(2)}(\chi)$ with the energy of thermal activation ΔE_1. The second phase includes HA tunneling in the energy-equivalent state $\psi^{(2)}(\chi')$ of the subsystem {3}, which formed due to thermal fluctuations. It is assumed that the wells in the T-positions χ and χ' are symmetrical. The contribution of the second phase ΔE_2 to the total activation energy is equal to the minimum work $\tilde{w}_{\lambda'\lambda}$, which must be performed in order to transfer the subsystem {3} from the state without deformation λ to the state with deformation λ'.

In the conditions of the thermally activated process the tunneling matrix element can not be approximated by (8.15). We use the definition applicable to the transitions of the HA between excited states of two harmonic wells [197,198]

$$J = \omega^2 \left(\frac{m\omega_2 d^2}{\pi} \right)^{1/2} e^{-m\omega_2 d^2/4} \cdot G_{\lambda'}, \qquad (8.23)$$

where $\omega_2 = \tilde{\varepsilon}_2 - \tilde{\varepsilon}_1$, $G_{\lambda'}$ is the contribution to local deformation. Moreover, the value $G_{\lambda'}$ can be computed in the one-dimensional model. Indeed, consider a crystal lattice as a set of linear chains of ions, oriented in the direction T–T. We select one of them – a linear chain of the T–T transition where the HA is located in the interstitial position. In this case, it suffices to take into account the displacement of ions of the given chain, and the deformation of neighboring chains in the first approximation is not considered.

The additional energy of the system due to displacements of the ions is described by a Hamiltonian which is linear with respect to the operators of the phonon field b_k^+ and b_k. Therefore, the contribution of deformation to the velocity operator of the HA can be distinguished as a separate factor by using a unitary transformation $\tilde{V} = e^{-S} V e^{S}$, where S is also the linear operator with respect to b_k^+ and b_k:

$$S = \sum_m S(\tilde{x}_{m-1}) a_m^+ a_m, \quad S(x) = \frac{1}{\sqrt{\tilde{N}}} \sum_k \{u_k(x) b_k^+ - u_k^*(x) b_k\},$$

where a_m^+ and a_m are the creation and annihilation operators of the HA at site $\tilde{x}_m$. We get

$$\tilde{V} = \langle 0|v|d \rangle \sum_m \Phi_m a_{m+1}^+ a_m,$$

where the factor Φ_m determines the contribution of motion of the deformation cloud. Let f_k denote the Fourier component of the force acting from the side of the HA, then

$$\Phi_m = \exp\left\{ \frac{1}{\sqrt{\tilde{N}}} \sum_k \left[b_k^+ \Delta_k(m, m-1) - b_k \Delta_k^*(m, m-1) \right] \right\},$$

$$\Delta_k(m, m-1) = u_k(\tilde{x}_m) - u_k(\tilde{x}_{m-1}),$$

where $u_k(\tilde{x}_m) = u_k e^{-ik\tilde{x}m}$ is the dimensionless displacement of the ion, $u_k = -if_k (2M\omega_k^3)^{-1/2}$.

As the definition of the diffusion coefficient we use the following expression

$$\mathcal{D} = \int_0^\infty dt \, \mathrm{Re} \langle \tilde{V} \tilde{V} + (-t) \rangle e^{-\tilde{\gamma} t}.$$

Given the explicit form of $\tilde{V}$, we obtain $\mathcal{D} = D^{(d)} + D^{(n)}$ where

$$\mathcal{D}^{(d)} = \left| \langle 0|v|d \rangle \right|^2 \sum_{m m'} \int_0^\infty dt \, \mathrm{Re} \left\{ \langle \rho_{m'}^+ \rho_m(-t) \rangle \langle \Phi_{m'} \rangle \langle \Phi_m^+ \rangle \right\} e^{-\tilde{\gamma} t},$$

$$\mathcal{D}^{(n)} = \left|\langle 0|v|d\rangle\right|^2 \sum_{mm'} \int_0^\infty dt \, \mathrm{Re}\left\{\langle \rho_{m'}^+ \rho_m(-t)\rangle \langle\left(\Phi_{m'} - \langle\Phi_{m'}^+\rangle\right)\rangle \times\right.$$

$$\left.\times \langle\left(\Phi_m^+(-t) - \langle\Phi_m^+\rangle\right)\rangle\right\} e^{-\tilde{\gamma}t}, \quad \tilde{\gamma} \to +0,$$

where $\rho_m = a_{m+1}^+ a_m$. Term $\mathcal{D}^{(d)}$ corresponds to the 'diagonal' transitions of the HA in which the occupation numbers of phonons do not change. Such transitions can be implemented at very low temperatures. Term $\mathcal{D}^{(n)}$ corresponds to the 'non-diagonal' transitions accompanied by absorption and emission of phonons. The change of the occupation numbers in such transitions is due to the fact that the restructuring of the oscillations of the chain relative to new equilibrium positions of the ions does not have time to occur fully during hopping of the HA. In the phonon-induced transitions, which are actually thermally activated tunneling processes, this term is of primary importance.

We introduce into $\mathcal{D}^{(n)}$ the spectral intensity $J_{mm'}$ of the correlation function $\langle \rho_{m'}^+ \rho_m(-t)\rangle$. The result is

$$\mathcal{D}^{(n)} = \left|\langle 0|v|d\rangle\right|^2 \sum_{m'm} \int_{-\infty}^\infty d\omega \, J_{mm'}(\omega) K_{mm'}(\omega),$$

$$K_{mm'}(\omega) = \int_0^\infty dt \, e^{i(\omega+i\tilde{\gamma})t} \left\{\langle \Phi_{m'}\Phi_m^+(-t)\rangle - \langle\Phi_{m'}\rangle\langle\Phi_m^+\rangle\right\}. \tag{8.24}$$

The analysis conducted in [199] showed that the main contribution to the diffusion coefficient comes from small values of ω, more precisely, the values of $\tilde{\omega}$, for which $\beta_1\tilde{\omega} \ll 1$ and $\tilde{\omega}t_0 \ll 1$, where t_0 is the time during which the integrand in $K_{mm'}(\omega)$ drops sharply. Then $K_{mm'}(\omega)$ in (8.24) can be replaced by $K(0)$ (here we omit the indices m and m', since the integral is calculated in the approximation of nearest neighbors, and in this approximation it is independent of the coordinates of the nodes). After this replacement, the dimensionless quantity $G_{\lambda'}^2 = K(0)\left(d/2\langle v\rangle\right)^{-1}$, appearing in the diffusion coefficient, is determined by the lagging of movement of the deformation cloud accompanying motion of HA. In other words, $G_{\lambda'}^2$ is associated with a delay of the so-called 'fur coat' [195] on the motion of the HA.

Using the expression for the diffusion coefficient (8.24) and using (8.23) and equation $\mathcal{D} = \Gamma R^{(a)} \sim G_{\lambda'}^2$, we find

$$G_{\lambda'}^2 = K_{\lambda'} \left(\frac{d}{2\langle v \rangle} \right)^{-1},$$

$$K_{\lambda'} = \int_0^\infty dt \left\{ \left\langle \Phi_m \Phi_m^+ (-t) \right\rangle - \left\langle \Phi_{m'} \right\rangle \left\langle \Phi_m \right\rangle \right\} e^{-\bar{\gamma} t}.$$

Here Γ is a geometrical factor, $\langle ... \rangle$ is the statistical average. K_λ is computed in the approximation of the nearest neighbors. We get

$$K_{\lambda'} = e^{-2S_0} \int_0^\infty dt \, e^{-\bar{\gamma} t} \left\{ \exp\left[2\sum_k |u_k|^2 (1 - \cos kd) \cos(\omega_k t) \times \right.\right.$$

$$\left.\left. \times \left(\operatorname{sh}\left(\frac{\beta_1 \omega_k}{2} \right) + \frac{\beta_1 \omega_k}{2} \right)^{-1} \right] - 1 \right\},$$

$$S_0 = \sum_k |u_k^2| (1 - \cos kd) \operatorname{cth}\left(\frac{\beta_1 \omega_k}{2} \right).$$

The integration over dt in $K_{\lambda'}$ is carried out by the method of steepest descent. But the greatest contribution comes from the first saddle point. In this approximation

$$K_{\lambda'} = \frac{\sqrt{\pi}}{2} \tau_r e^{-\beta_1 \Delta E_2},$$

$$\left(\frac{1}{\tau_r} \right)^2 = \frac{1}{\tilde{N}} \sum_k |u_k|^2 (1 - \cos kd) \frac{\omega_k^2}{\operatorname{sh}(\beta_1 \omega_k / 2)}, \qquad (8.25)$$

$$\Delta E_2 = \frac{1}{\tilde{N}} \sum_k |u_k|^2 \omega_k (1 - \cos kd).$$

The total activation energy of the jump in the case of thermally activated tunneling is $E^{(a)} = \Delta E_1 + \Delta E_2$. To calculate ΔE_1 in the variational problem, we take as a test function of the excited state the hydrogen-like function (2S-state) with one node $\psi^{(2)} = \psi^{3/2} 2^{-3/2} \pi^{-1/2} e^{-\gamma R/2} (1 - \gamma R/2)$.

The selected function satisfies the normalization condition and is orthogonal to $\psi^{(1)}$. Substituting $\psi^{(2)}$ in (8.19), we obtain an expression for the functional $\mathcal{E}_{\lambda'}\left[\psi^{(2)} \right]$ to investigate its extreme values. Without citing here cumbersome expressions, we note that the functional $\mathcal{E}_{\lambda'}\left[\psi^{(2)} \right]$ has a minimum, which corresponds to the stationary excited state of the subsystem $\{2\}$. Calculation of ΔE_1 with the variation of elastic deformation energy (8.18) taken into account is reduced to minimize the difference between the functionals (8.19)

$$\Delta E_1 = \min\left\{ \mathcal{E}_{\lambda'}\left[\psi^{(2)}\right] - \mathcal{E}_{\lambda'}\left[\psi^{(1)}\right]\right\}. \qquad (8.26)$$

The value ΔE_2 can also be calculated within the framework of the variational problem. Indeed, as noted above, $\Delta E_2 = \tilde{w}_{\lambda'\lambda}$; the latter is calculated by analogy with the calculation of $\tilde{w}_{\lambda'\lambda}$, from (8.21).

Now, instead of J_{23}, we introduce the rate constant for the matrix element defined by (8.23)–(8.25) and replace ζ by $\beta_1\Delta E_1$. The result is the rate of thermally activated tunneling of the HA

$$R^{(a)} = \frac{1}{2\sqrt{\pi}}B(m)\beta_1^{1/2}\Delta E_1 e^{-\beta_1 E^{(a)}}\left\{\frac{\tau_r}{\tau_1}\tilde{\varphi}\left(\beta_1\,\omega_2\right)\right\}, \qquad (8.27)$$

where $B(m) = m^{1/2}\omega_2 de^{-m\omega_2 d^2/2}$.

We make two additional observations. First, the real values $\zeta = \beta_1\Delta E_1$ lie in the range $2 < \zeta < 5$, so if we write the expression (8.27) it is more accurate to take η from (8.10). Then an additional factor $1 + \mathrm{Ker}\,(\beta_1\Delta E_1/2.2)$ appears in (8.27), in which the second term gives a small correction to the first. Secondly, the sub-barrier transition of the HA with the rate coefficient (8.27) in the classification of [199] should be attributed to the number of 'active' jumps. The transition (8.22) can be considered 'passive'. Recall that $E^{(p)}$, appearing in (8.22), is the activation energy of the lattice, rather than the activation energy of the migrating atom.

In the 'frozen' well V_1 there are two levels: ε_1 – the level of the ground state, self-consistent with the potential V_1, and $\tilde{\varepsilon}_2$ – the level of the excited state, with non-self-consistent V_1. The transitions of the HA between the levels of the 'frozen' well occur instantaneously and can be implemented in quasi-elastic scattering of slow neutrons on the HA. The restructuring of local deformation and the formation of the self-consistent excited states whose energy is lower than $\tilde{\varepsilon}_2$, occurs after the thermal excitation of the HA.

In the first phase of the hopping (Fig. 8.5a) the HA is transferred from the ground state to an excited state. Excitation of the atom with the thermal activation energy ΔE_1 is accompanied by relaxation of the lattice, which in turn leads to the formation of a new potential well V_2 and changes the nature of the oscillations of the HA, so that thermal excitation is followed by the formation of an asymmetric two-cell well, including a well V_2 in the T-position and V_3 in T'-position (Fig. 8.5b). In the second stage of the active transport process is tunneling (sub-barrier) transition of an atom from a given T-position to the nearest neighbor T'-position, followed by

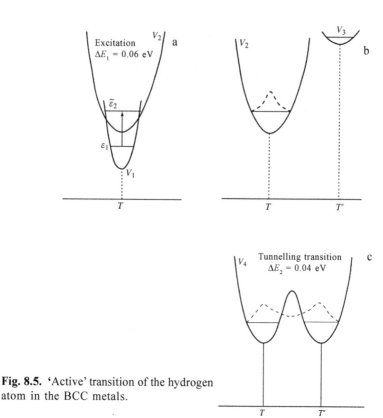

Fig. 8.5. 'Active' transition of the hydrogen atom in the BCC metals.

the movement of the deformation cloud. The 'polaron' effects in the strong coupling conditions give an additional contribution ΔE_2 to the total activation energy of the process as a whole.

Using representations similar to the classical diffusion theory, the transition to a neighboring tetrahedral position can be interpreted as a transition of the HA to the next (energy equivalent) state of the symmetrical well V_4, arising from thermal fluctuations (Fig. 8.5c). The additional energy for the formation of such a fluctuation ΔE_2 is realized through the elastic energy that is released by resorption of the deeper potential well V_1 of the ground state. The total activation energy of thermally activated tunneling of HA is equal to $\Delta E^{(a)} = \Delta E_1 + \Delta E_2$. In the special case of HA in niobium theoretical values are: $\Delta E_1 = 6 \cdot 10^{-2}$ eV, $\Delta E_2 = 4 \cdot 10^{-2}$ eV. The result of the theory $\Delta E^{(a)} = 10^{-1}$ eV, is in good agreement with experimental data $E^a_{\exp} = 1.06 \cdot 10^{-1}$ eV [200].

8.4. Concluding remarks on the migration of hydrogen atoms

8.4.1. Thermal fracture of activation energy near Debye temperature θ_D

The physical picture of the formation of a quantum state, as well as changes in the activation energy of the HA, were discussed in several papers, in particular, in [201]. The theory is based on the band structure of sinusoidal potentials. Moreover, the complex picture of the temperature dependence of diffusion was associated with the broadening of energy levels (bands) in relation to temperature. At the same time, the zone of the ground state was considered so narrow that any distortion of the lattice led to its destruction and, therefore, the localization of the impurity center. Note that to interpret such complex phenomena as the activation energy of diffusion, it is necessary to use the potential of a more adequate reality than the simplified sinusoidal potential, namely the potential which significantly change when the form of the wave function of the HA changes.

Another interpretation of the diffusion migration of the HA is proposed in this book. In fcc metals there are above-barrier jumps, since the deep potential wells are due to the effect of Coulomb forces. In bcc metals, the formation of deep wells is associated with the deformation effects, and two transfer channels with the rate constant, defined by (8.22) and (8.27), appear. The results of numerical calculation of the activation energy of 'passive' and 'active' transport will be discussed to identify the mechanism of the elementary act in different temperature ranges. In the case of diffusion of the HA in niobium calculations $E^{(p)}$ and $E^{(a)} = \Delta E_1 + \Delta E_2$ give 0.07 eV and 0.10 eV, respectively.

Consequently, the theoretical considerations made in this book can be used to explain the result which shows that a transition from the activation energy of 0.068 eV to 0.106 eV takes place in the vicinity of Debye temperature [202]. We attribute this fact to the transition from 'passive' transfer of HA at $\beta_1^{-1} < \theta_D$ to 'active' $\beta_1^{-1} < \theta_D$. Thus, the temperature inflection of activation energy (Fig. 8.6) is caused not by the change in the efficiency of channels (bands) of the same 'frozen' well, as suggested in [201], but by the change of the elementary act of diffusion of the HA.

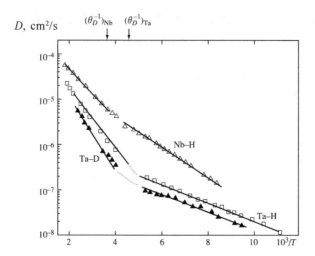

Fig. 8.6. Coefficients of diffusion of hydrogen and deuterium in niobium and tantalum. (θ_D = 277 K (Nb), 240 K (Ta)).

8.4.2. Classical and anomalous isotope effects

Migration of HA in fcc metals is accompanied by the classical isotope effect. Indeed, taking into account (8.12), as well as the dependence of $\Omega \sim m^{-1/2}$ at high temperatures, we have

$$R^{(f)} \sim \frac{1}{m}\tilde{\varphi}(\xi) \sim \frac{1}{m\Omega} \sim m^{-1/2}, \quad \mathcal{D} = \Gamma R^{(f)} \sim m^{-1/2}.$$

This isotopic dependence of the diffusion coefficient $\mathcal{D} \sim m^{-1/2}$ has been repeatedly observed. For comparison, we present the results of other theories: $\mathcal{D} \sim m^{-2}$ [82], $D \sim m^{-1}$ [186].

Using (8.12), the isotopic ratio for the HA and its isotopes in fcc metals can be written as

$$\frac{\mathcal{D}(m)}{\mathcal{D}(2m)} = \begin{cases} 2e^{-0.29\beta_1\Omega} & \text{for } \beta_1\Omega>1, \\ \sqrt{2} & \text{for } \beta_1\Omega \ll 1. \end{cases}$$

The temperature dependence of this ratio is shown in Fig. 8.7 (see curve 1 for the system (Cu–H)). When $\beta_1^{-1} < \theta_D$ a significant contribution to the diffusion of the HA is provided by quantum statistical effects, because in this temperature range the thermal de Broglie wavelength for the HA is comparable with the lattice constant. Due to these effects, the ratio of $\mathcal{D}(m)/\mathcal{D}(2m)$ is significantly smaller than its classical value of $\sqrt{2}$. However, with

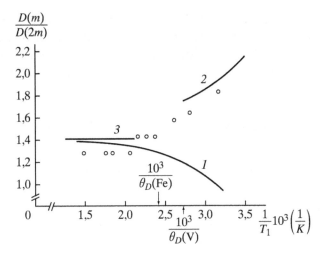

Fig. 8.7. Normal (curve 1) and anomalous (lines 2, 3) isotope effect.

increasing temperature, as shown in Fig. 8.7, the isotopic ratio increases monotonically, approaching $\sqrt{2}$.

Another isotopic dependence of the ratio $D(m)/D(2m)$ on temperature is observed in the diffusion in bcc metals. In 'passive' tunnel transition, $\beta_1^{-1} < \theta_D$, the minimum work $w_{\lambda\lambda'} = E^{(p)}$ (8.21) depends on the mass of the isotope (see Fig. 8.3), and according to (8.22), the isotopic ratio becomes

$$\frac{D(m)}{D(2m)} = C_1 e^{\beta_1 \left(w_{\lambda\lambda'}(2m) - w_{\lambda\lambda'}(m) \right)}.$$

Since C_1 is a constant factor, then with the inequality $E^{(p)}(\gamma_D) > E^{(p)}(\gamma_H)$ taken into account, the ratio $D(m)/D(2m)$ decreases with increasing temperature, as illustrated by curve 2 in Fig. 8.7.

In the case of 'active' tunneling, $\beta_1^{-1} < \theta_D$, formula (8.26) gives

$$\frac{D(m)}{D(2m)} = 2 \frac{B(m)}{B(2m)} \frac{e^{\beta_1 \omega_2 / \sqrt{2}} - 1}{e^{\beta_1 \omega_2} - 1} = \sqrt{2} e^{C_2 m \omega_2 d^2},$$

where the exponential multiplier on the right side of this equation is close to unity, as $C_2 \approx 0.1$. Consequently, when $\beta_1^{-1} \gg \theta_D$, the ratio $D(m)/D(2m) \approx \sqrt{2}$, as shown by straight line 3 in Fig. 8.7. The lines 2 and 3 are in qualitative agreement with the temperature course of the relationship $D(m)/D(2m)$, which is observed (see Fig.

8.3) for the diffusion of HA and its isotopes in bcc metals [203]. If the function $\mathcal{D}(m)/\mathcal{D}(2m)$ in above-barrier jumps is close to the classical value $\sqrt{2}$ on the side of smaller values, then in the case of tunneling it approaches from large values and this appears as the anomalous isotope effect.

8.4.3. Departure from the concept of hopping conduction of small radius polarons

Initially, the theory of thermally activated processes of tunneling in solids has been developed for small-radius polarons in ionic crystals [204]. This was then extended to the case of hopping diffusion in of HA in metals [192–195,205,206]. In these studies, attention was paid to the tunneling of HA in both fcc and bcc metals, and the contribution of local deformation provided by the impurity center was taken into account.

However, from the very beginning of development of the theory it was clear that there are other, more complex elementary processes by which diffusion migration of the HA is possible. In addition, it was found that the theory based on the concept of polarons can not describe the diffusion of hydrogen in bcc metals at temperatures above 250 K [207]. More precisely, it can not give the observed isotopic dependence of the diffusion coefficient of the HA. In addition, because of the very small value of the of tunneling matrix elements doubts were expressed as to the applicability of the theory of small polarons to the problems of hydrogen diffusion in fcc metals.

A detailed analysis of the situation at the beginning of the 1990s in the diffusion of hydrogen and also of contradictions associated with the analogy of migration of the HA and movement of the polaron, is presented in a literary review by the author of this book [190]. Taking into account these difficulties, we did not use the concept of conduction of small polarons and used instead the ideas of Prigogine and Beck [82]. So the basic relations (8.22) and (8.24) were obtained on an entirely different basis.

As for the publications of the last decade, the author does not know of any work discussing elementary diffusion act of the HA by modern statistical thermodynamics.

Conclusion

Basic equations of non-equilibrium statistical thermodynamics of fast particles moving in solids

The new direction in the velocity theory is based on the description of irreversible processes in the framework of statistical Gibbs ensembles for non-equilibrium systems. For systems that include fast particles as one of the subsystems, the book describes a new approach to deriving equations of the relaxation processes that are non-linear with respect to thermodynamic forces. The non-linearity is due to the fact that the thermodynamic parameters were included in the kinetic coefficient with a considerable delay in different subsystems, with the exception of thermodynamic forces. A generalized expression was derived for the fundamental kinetic coefficient – the rate coefficient as a product of the spectral intensities of the time correlation functions. In contrast to the rate constant in the theory of absolute velocities, the generalized expression determines the rate of processes in a wide range of thermodynamic conditions.

It has been established that the two-time density–density Green function, which is calculated using the spectral intensity, satisfies the finite-difference equation of the Fokker–Planck–Kolmogorov type, which describes both the coherent nature of the particles in a regular crystal and inelastic scattering processes on electrons and thermal vibrations of the lattice atoms. The main transport equation of the theory has an analytical solution for a class of degenerate hypergeometric functions $_1F_1(\alpha, \gamma; x)$ for $\gamma = 1$, and in the same class of functions the generalized expression was derived for the velocity of the particles from the potential minimum.

It is important to note a characteristic feature of our proposed approach: damping of the main function, i.e. the density–density Green function is determined by a superposition of exponential functions with a certain set of relaxation constants, which corresponds to the asymptotic form of time correlation functions in the stochastic case. All statistical averaging is performed by the non-equilibrium statistical operator; the statistical properties of the correlation functions determined by density fluctuations in the number of particles, as well as the correlation functions of fluctuations of the forces and fluxes.

The effectiveness of the developed formalism was demonstrated on the example of studies of orientation effects in the crystals. Orientation effects are observed during the passage of high-energy charged particles – protons, light atomic particles, electrons – through the crystals. They are due to the fact that the character of the movement and interaction of particles with lattice atoms depends strongly on the orientation of the directional beam of fast particles in relation to the main crystallographic axes and planes of the crystal. Theoretical studies of orientation effects and, in particular, of the channeling effect include studies of aspects which gives a generalized expression for the rate coefficient: the rate coefficient of diffusion dechanneling in a rigid lattice in electron scattering, and the rate coefficient in phonon scattering.

Of course, the study was carried out under the assumption that in the system there is always a significant difference between the characteristic times of the collisions and the time scales of relaxation processes. Because of this relationship complicated transition processes (the smoothing process) decay quickly enough and the kinetic equations, as well as stochastic equations of motion, are a reliable approximation for the analysis of time evolution of a beam of fast particles.

It is established that in the semi-classical approximation, the coefficient of the rate of dechanneling includes, in addition to the factors known from the phenomenological theory, as well as the total cross section, two additional functions $k(\zeta)$ and $v(F_{21})$. The first of these functions describes the kinetic regularities for the disappearance of the channeled particles from a potential minimum (the channel), and second is determined by separation of the thermodynamic parameters of the subsystem of channeled particles from the parameters of the thermostat. As for the total cross section of electron scattering, it corresponds to the total cross section of the van Hove theory which investigates the scattering in a many-electron system of a more general type. The total cross section for the interaction of particles with the lattice vibrations is determined by one-phonon and two-phonon scattering processes.

The high efficiency of the thermodynamic approach is shown in the case of transfer processes accompanying the motion of channeled and random particles in solids. The thermodynamic relaxation of positronium atoms and cascade particles in solids, as well as hot carriers in semiconductors, were studied; the transition to fast particle moving with thermal velocities was performed. In addition, the important result of the theory described in this book should be regarded as proof of the positive sign of the intensity of the source of entropy, which characterizes the local entropy production (sections 3.5, 4.1). The second law of thermodynamics is nothing more than a statement about the positive sign of the entropy of the source. Implementation of the second law provides a basis for fast particles to be considered as an independent thermodynamic subsystem. The kinetic equation (3.15)–(3.16), constructed on the basis of the BBGKY chain and the relations of classical field theory, has allowed us to reveal the time evolution of entropy in the case of a spatially inhomogeneous system. The presence of spatial heterogeneity required the use of a completely different approach than the direct substitution of the kinetic equation into the expression for the derivative of the entropy with respect to time.

Temporal evolution of particles of a directional beam

We confine ourselves to planar channeling. In general, the thermodynamic state of the system, which includes the fast particles as one of the subsystems, is determined, on the one hand, by elastic collisions, which lead to a redistribution of energy with respect to translational degrees of freedom of the particles, on the other hand - by inelastic collisions leading to excitation of vibrational degrees of freedom.

Inelastic collisions result in a continuous exchange of energy between translational and vibrational degrees of freedom, which in the case of channeling is accompanied by an increase in the transverse energy of channeled particles (CP). An increase in the transverse (vibrational) energy of the particles, of course, is accompanied by a slight retardation. Also, the elastic collisions are most effective in small depths (section 2.5); of course, they also occur at great depths in excess of the coherence length. However, at great depths their role is quite different. At great depths elastic collisions are responsible mainly for the dynamic effects of the collision [149] and are essentially the adiabatic reaction of the subsystem of particles in interaction with the thermostat.

Let us consider the temporal evolution of the CP from the viewpoint of classical theory. When considering the effect of channeling, there are two stages:

– The first stage – a stage of smoothing in which the channeling mode forms. In the classical subsystem the CP subsystem is described by $\sigma(x,p_x,z)$ – the joint probability density with respect to the transverse coordinates and transverse momenta. At shallow depths, the passage of particles through a crystal on which the smoothing process takes place inelastic scattering still not efficient enough. Under these conditions, the potential disharmony of the planar channel comes to the fore and also leads to a smoothing of the spatial probability density and to the smoothing in the space of transverse momentum with increasing penetration depth into the crystal. (The smoothing of a classical channeling picture as a result of anharmonism of vibrations of the particles can be compared with attenuation of the off-diagonal density matrix elements in the quantum picture of the effect.) The first stage ends with the formation of the particle distribution function, which depends on the transverse energy $\varepsilon_\perp = \left(p_x^2/2m\right)+\left(1/2\right)m\Omega^2 x^2$.

From the standpoint of the analytical theory, the formation of such a function can be described as follows. We assume that the interaction of CP with the thermostat is proportional to the small parameter α. We choose a time interval Δt so that, on the one hand, Δt is much larger than the period of oscillations in the CP channel and on the other hand – the effect of the thermostat on the CP during Δt can be considered quite small. The probability density $\sigma(x, p_x, z)$ is then transformed into $\sigma(\varepsilon_\perp, t)$ in the sense that

$$\frac{1}{\Delta t}\int_t^{t+\Delta t} dt\, \sigma\left(x,p_x,t\right) \to \sigma\left(\varepsilon_\perp,t\right),$$

i.e. there is uniform convergence of functions $\sigma(x, p_x, t)$ to $\sigma(\varepsilon_\perp, t)$ for any sequence of Δt.

– The second stage of evolution is the relaxation stage. After smoothing, as the analysis shows, the probability density of the $\varepsilon_\perp$-space in a sufficiently large time interval can be approximated by the function

$$\sigma\left(\varepsilon_\perp,t\right)=G\left(\varepsilon_\perp\right)\frac{1}{2\pi}\lambda\int_0^{2\pi} d\theta \exp\left\{-F_{21}\left(t\right)\times\right.$$
$$\left.\times\left[\varepsilon_\perp - \varepsilon_m \exp\left(-2\alpha^2\gamma t\right)-2\sqrt{\varepsilon_\perp\varepsilon_m}\exp\left(-\alpha^2\gamma t\right)\cos\theta\right]\right\}.$$

Here, as usual, λ is the normalization factor, $G(\varepsilon_{\perp})$ is the number density of states of the CP, $1/F_{21}(t)$ is the current value of the transverse quuasi-temperature. In addition, $\sigma(\varepsilon_{\perp},t)$ includes the dimensionless square matrix element of the interaction of CP with the thermostat γ, and the relationship of parameter α with the sequence limit Δt is defined as follows:

$$\lim \alpha^2 \Delta t = 0 \qquad \text{and} \qquad \alpha \to 0.$$

The resultant density in the one-particle density distribution of the CP in a non-equilibrium state at time t, which completely preserves the 'memory' of the initial state particles or, more accurately, of the transverse energy of entry of the particles into the channel ε_m. Naturally, in transition to a quasi-equilibrium, in other words, in the limit $t \to \infty$ this function 'loses memory' of the initial state of CP, which is consistent with the general concept of non-equilibrium statistical mechanics.

It is well know that relaxation phenomena are studied using kinetic equations. If in the resulting expression for $\sigma(\varepsilon_{\perp},t)$ the function $G(\varepsilon_{\perp})$ is approximate by a continuous step function, the smoothed function satisfies the Fokker–Planck equation. The same physical situation is found in the quantum theory of channeling. Indeed, after the decay of non-diagonal density matrix elements the temporal development of the diagonal matrix is described by the equation of the Fokker–Planck type.

Anharmonicity of oscillations of particles of a directional beam

It has been established that the anharmonicity of the CP between the channel walls has a significant influence on the temporal evolution of the spatial and angular distributions of particles of the directional beam. However, the role of anharmonicity in the orientational effects is much wider. For example, the equilibrium quasi-temperature (6.26) is established not only through the diffusion of CP in the space of transverse energies but also dissipative process.

The exchange of energy between the CPs also plays an important role and leads to the redistribution of energy over the vibrational degrees of freedom within the subsystem of the particles of the directional beam. In other words, the exchange determines the internal thermalization of the subsystem of the CPs. It is known that the internal redistribution is due to the influence of anharmonic effects [51] which enable transitions of particles from a state with a given frequency to a state with a different frequency.

These transitions are associated with certain acts of 'collisions'. In the case where the anharmonic term is less than the harmonic one, the result of 'collisions' is calculated by the perturbation theory. Of course, the time between 'collisions' computed in this way contains the coupling parameters of the same order as the anharmonic term treated as a perturbation. For example, the inverse lifetime of the local mode, which is determined by the acts of 'collisions', $1/\tau_s$, is proportional to the product of two coupling parameters of the fourth order.

The combined impact of two scattering factors leads to the appearance in the relaxation equations of additional terms describing the dynamical effects of collisions in terms of interference of the types of interaction [208]. We have already encountered a similar situation when studying the spin quasi-temperature, but there we considered the interference of the spin–spin and spin–lattice interactions. In contrast to the spin theory, the problem of channeling is concerned with the interference of electron scattering and anharmonic effects.

Taking into account the similarity of physical situations, as well as the coincidence of the structure of the expressions for CP (6.26) and nuclear spins (6.52), it is enough to use for expressing the interference contribution in (6.26) by the transformation, which has previously been used in the theory of spin quasi-temperature. Namely, we replace the value of B in the denominator of the second term in (6.26) by $B\eta_c$. The explicit form of the factor is determined from the relaxation equation, which takes into account the contribution of the fourth-order anharmonicity,

$$\eta_c = \mathrm{ch}\left\{\left(\bar{\omega}\tau_{so}\right)^{-1/2}\right\}.$$

When writing the renormalization factor we use the notation: $\bar{\omega}$ is the frequency of oscillation of the CP in the channel in the harmonic approximation, $\bar{\omega}s_0$ is the height of the potential barrier of the channel wall. After replacing $B \to B\eta_c$ in (6.26) the expression for the transverse quasi-temperature of the CP takes its final form

$$\frac{1}{F_{21}} = \frac{1}{\beta_1}\left(1 + \frac{A}{B\eta_c}\right).$$

The final value of quasi-temperature is formed with the participation of three processes:

1) diffusion of CP in the space of transverse energy ('heating' of CP);

2) a dissipative process due to the drag of electrons by the oscillating scattering center, i.e. CP ('cooling');

3) rapid internal thermalization of the CP subsystem due to the anharmonicity of vibrations of particles, which leads to energy exchange between the CP.

Temporal evolution and thermodynamics of the chaotic particle beam

The temporal development of the chaotic particle beam will be discussed on the example of the evolution of the subsystem of knock-on atoms (KOA) formed in displacement cascade. In this case, the KOA reproduction takes place in the first (dynamic $\sim 10^{-13}$ s) stage of the cascade process. Reproduction occurs as a result of elastic collisions of the PKOA with atoms of the crystal.

The next (relaxation) stage is characterized by the redistribution of energy derived from the radiation by subthreshold collisions. It is at this stage where the subsystem of the KOA 'heats up' and the quasi-equilibrium state of the KOA forms, i.e. non-equilibrium steady state, far away from thermodynamic equilibrium with the lattice. Since the time evolution of the cascade at the relaxation stage can be considered as a Gaussian–Markov process, the time dependence of the quadratic correlator which has entered into the formula for the quasi-temperature of the KOA, can be approximated by a function that is characteristic for Gaussian fluctuations, with relaxation time τ_R. In this case, the time course of the quasi-temperature is heating is given by

$$\frac{1}{F_{21}(t)} = \frac{1}{F_{21}}\left(1 - \frac{1}{2}e^{-2t/\tau_R}\right)$$

and the establishment of quasi-equilibrium in the subsystem of the KOA corresponds to the limit $t \to \infty$. In a real physical situation, the stationary quasi-temperature is established at a time when the rate of energy intake is balanced by radiation losses in the heat transfer.

No less interesting stage of development is the chaotic thermalization of the beam. (Thermalization can not take place in the case for a directional beam: particles initially come from the regime of directed motion, and only then are thermalized). Transition to the full thermodynamic equilibrium of the whole system is considered as a transition through several intermediate states of quasi-equilibrium with the current value of quasi-temperature.

Evolution of quasi-temperature during thermalization is described by a kinetic equation for quasi-temperature (as opposed to the particle distribution function). This equation includes the contribution of energy dissipation due to the interaction of KOA with thermal lattice vibrations and the contribution to the energy flux due to the recombination of the unstable defects. The thermalization problem is formulated in the framework of non-equilibrium statistical thermodynamics, which leads to a non-trivial result. It is shown that the decrease in the quasi-temperature of the KOA is described by a linear function of time, if we consider the early period of thermalization, and at later stages the rate of decrease is lower.

More complex versions of the thermalization of chaotic particles are also found. For example, the thermalization of hot carriers in polar semiconductors in the laser field is divided into two periods. In the first period carriers transmit extra energy to the subsystem of optical phonons, which leads to internal thermalization of the two subsystems. Only at the next stage the two subsystems transfer the energy to the thermostat – acoustic phonons. Therefore, the non-equilibrium steady state reached in the initial period is considered as an initial condition for the thermodynamic relaxation of carriers and the thermostat.

Kinetic and thermodynamic characteristics of systems in conditions of diffusion processes

The theory of absolute reaction rates is applicable to systems close to thermodynamic equilibrium. In contrast to this situation, non-equilibrium thermodynamics applies to a wide class of phenomena, and this helps to consider many diffusion processes from the same viewpoint. As shown above, the basis of the diffusion processes which we examined are random walks of particles between levels of multilevel wells having reflecting and absorbing barriers.

However, the thermodynamic states of systems considered in problems of diffusion dechanneling and migration of light impurity particles are fundamentally different: in the case of channeling the system consists of two non-Boltzmann subsystems and the thermostat and (of course, the Boltzmann subsystem), and in case of migration all the subsystems are of the Boltzmann type. This difference is reflected in the kinetics.

We discuss this issue in more detail by considering the problem of diffusion of particles from the potential minimum. Since the configuration impurity relief (to be specific, hydrogen atom in an fcc lattice) can be approximated by a symmetrical two-cell well, the ground state – levels of the left well, i.e. subsystem {2}, and right – subsystem {3} are energetically the same. The Boltzmann distribution of impurities $p(\varepsilon)$ is cut off at high energies ($\varepsilon \geq \overset{*}{\varepsilon}$) (Fig. Ia).

The quasi-equilibrium distribution of the CP is significantly different. Although the function $n(\varepsilon_{\perp})$ is an exponential distribution, the temperature of the thermostat is not its modulus. An important role in the formation of the full distribution $p(\varepsilon_{\perp}) = G(\varepsilon_{\perp}) \, n(\varepsilon_{\perp})$ is played by the rapid decline in the density of states $G(\varepsilon_{\perp})$, which can be regarded as approximately a step function. This function reduces the proportion of the high-energy part of the distribution of the CP and determines the maximum value of the distribution in the region of the transverse energy of entry of the particles ε_m into the channel. So in the channeling theory the complete system consists of two different subsystems, namely, the Boltzmann subsystem {1} (thermostat) and non-Boltzmann subsystem – subsystem of particles {2} (Fig. Ib).

As for the energy distribution of the system which includes, as one of the subsystems, a bundle of random particles, we merely refer to the distribution of cascade particles in metals. The latter, as is well known (Fig. Ic), agrees qualitatively with the distribution of molecules of the classical gas in the problem of the leak of the gas jet into a volume, filled with a different equilibrium gas.

If the rate of the quasi-chemical reactions, which simulates the diffusion process in the space of transverse energies, is greater than the rate of establishment of thermodynamic equilibrium in the system, the system as a whole is in a state far removed from equilibrium. Therefore, the state of the system is described by the thermodynamic conditions

$$F_{21} = F_{31} \ll F_{11} = \beta_1, \quad \left|F_{22} - F_{32}\right| = F_{21}\left|\mu_2 - \mu_3\right| \gg 1. \qquad \text{(A)}$$

Such a system is implemented in the channeling of atomic particles. Indeed, the reaction time is represented by the duration

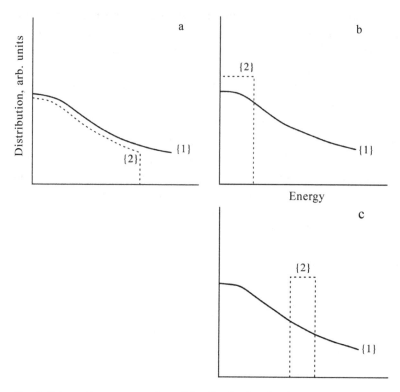

Fig. I. Schematic representation of the distribution for the subsystem {1} (thermostat) and the subsystem {2} (particles) in different physical situations. (Energy on the abscissa and the distribution on the ordinate are given in arbitrary units).

of motion of charged high-energy particles in the channel. (The latter is estimated as 10^{-12}–10^{-11}).

In the opposite case, when the rate of the simulated reaction simulation is lower than the rate of establishment of equilibrium, the system in terms of the diffusion process is close to thermodynamic equilibrium. This condition is typical for diffusion in configuration space, since the time of settled life of the migrating atom is of the order of 10^{-9}–10^{-8} s at a temperature of 10^3 K, which is several orders of magnitude greater than the characteristic time of channeling. Due to the exchange of energy over such a long duration of settled life the temperature of the subsystem of impurity particles is equalized with the temperature of the thermostat, so that we have approximately

$$F_{21} = F_{31} = \beta_1 \qquad\qquad (B)$$

The range of applicability of the common definition of the rate coefficient, formulated on the basis of the relations of non-equilibrium statistical thermodynamics, is very wide. It covers the processes that occur in the conditions (A), and those which correspond to the conditions (B). As an example, consider the four branches provided by the single definition of the rate of the diffusion process (Sections 5.3 and 8.1):

$$\text{Conditions (A).} \left[R \sim \frac{1}{F_{21}\omega_{so}} \left(\frac{1}{2} + \frac{1}{F_{21}\omega_{so}} \right)^{-1} \zeta j_{01}^{3/2} \exp\left(-\zeta j_{01}^2\right). \right.$$

$$\text{Conditions (B).} \left[\begin{array}{l} R^{(f)} \sim \dfrac{1}{\tau_1 \Omega} \beta_1 U_0 \exp\left(-\beta_1 U_0\right), \\[2ex] R^{(a)} \sim \dfrac{\tau_r}{\tau_1} \beta_1^{1/2} \Delta E_1 \exp\left(-\beta_1 E^{(a)}\right), \\[2ex] R^{(p)} \sim \dfrac{1}{\tau_1 \omega_1} \exp\left(-\beta_1 E^{(p)}\right). \end{array} \right.$$

The first branch corresponds to the rate of diffusion dechanneling R of the light atomic particles with energies of 1 MeV. As can be seen from the relationship, due to the small discreteness of the CP spectrum, diffusion of these particles in the space of transverse energies is a non-activated process. However, it should be noted that the kinetic coefficient R depends on the transverse quasi-temperature of the CP $1/F_{21}$. (Even more clearly, such a dependence is manifested in the equilibrium characteristics of channeling, see Section 6.5). This circumstance can be compared with the situation in the problem of dissociation of diatomic molecules in plasma. Dissociation, like diffusion dechanneling, occurs due to the rise of the molecules on the 'ladder' of vibrational states by transitions between levels with the result, as analysis shows, that the decay rate of molecules depends only on the vibrational temperature. (And there is a large gap between the vibrational and translational temperatures).

The second branch $R^{(f)}$ corresponds to the diffusion of HA in the fcc metals, which occurs in the thermodynamic conditions (B). Naturally, according to one degree of freedom – the number of particles, the HA subsystem is weakly non-equilibrium. The rate of migration in this case is determined by two factors: on the one hand, by the activation of the HA and transition T–O–O through the excited state, on the other hand – by the collisional deactivation of the atoms that have reached the height of the potential barrier.

The third and fourth branches $R^{(a)}$ and $R^{(p)}$ correspond to the diffusion migration of HA in the bcc metals, which is accompanied by movement or collapse of local lattice deformation. As shown, the lattice deformation in the vicinity of the HA leads to its self-capture. The interaction of HA with the strain field results in the formation of a deep potential well in which there is a level of the excited state.

In the first phase of thermally activated tunneling, the HA goes from the ground to an excited state with activation energy ΔE_1. At the second stage, tunneling transition takes place and is followed by motion of the deformation cloud, which gives an additional contribution ΔE_2 to the total activation energy $E^{(a)}$. These transitions in the bcc lattice are also accompanied by 'passive' transitions (the rate coefficient $R^{(p)}$), due to the collapse of deformation due to thermal fluctuations.

From a comparison of diffusion in of the HA in bcc and fcc metals it follows that the type of crystal lattice determines the type of mechanism of the elementary diffusion act. However, the temperature range in which migration occurs is also important. If there are several possible mechanisms, then the mechanism which operates with increasing temperature is the one which corresponds to a higher activation energy. As an example, data for the Nb–H system are given below.

Passive tunneling transitions	(120–220 K)	$E^{(p)} = 0.068$ eV
Activated tunneling transitions	(280–570 K)	$E^{(a)} = 0.106$ eV
Above-barrier jumps	(800–1100 K)	$U_0 = 0.144$ eV

Modes of motion of fast particles in the local theory

According to Lindhard's non-local classical theory, at angles of entry into the channel smaller than the critical channeling angle, the beam of fast particles is divided into two parts: channeled and chaotic. The channeled particles are moving in the center of the channel with periodic reflections from crystallographic planes forming the walls of the channel (oscillatory trajectories). In contrast, the chaotic particles move along straight paths crossing the plane at arbitrary angles.

Another picture of the orientation effects is given by the local theory regardless of what is its starting point: the local stochastic equations or the local Boltzmann kinetic equation. In the studies in this direction it was established that there is one more mode of particle motion in crystals – quasi-channeling. The fast particles, trapped in the quasi-channeling mode, move in the immediate vicinity of the crystallographic plane (or the axial channel wall) and travel to the adjacent wall in some random way.

Emphasis was placed on the consistent derivation of the local kinetic equations and the calculation, on the basis of these equations, of the equilibrium and kinetic characteristics of the system. In particular, the chain of BBGKY equations was used to obtain a local kinetic equation for the one-particle distribution function of fast particles, interacting with thermal lattice vibrations and valence electrons. Given the explicit form of the collision term which has entered into this equation, we find the kinetic function of the problem: the diffusion function in the space of transverse energies and the function that characterizes the energy loss of fast particles. Both kinetic equations are written in the form of an integral transform of the local matrix of random effects; the local theory approach should be considered as the direction of the theory of motion of atomic particles in crystals, additionally taking into account the transverse coordinates of the fast particles.

The functional dependence of the kinetic equations is different in the three regions of transverse energy. These areas correspond to the three modes of motion: channeling, quasi-channeling and chaotic motion. The kinetic functions have the following characteristics: diffusion function – break in the transition from channeling to quasi-channeling, and energy losses – the maximum that is associated with additional losses due to quasi-channeling. The fact is that in the quasi-channeling mode the particle moves near the plane where the intensity of its interaction with the vibrations of atoms of the plane is maximum. However, the density of the electrons in this region increases, which also leads to an increase in dynamic friction. Taking into account the peak loss, it can be argued that the local statistical theory proves the existence of particles with the inhibition stronger than the inhibition of an amorphous medium.

As far as the average number density of particles is concerned, the presence of particles, moving along trajectories of the oscillator, leads to the formation of an almost uniform density distribution along the transverse coordinate. On the other hand, the particles passing through the crystal in the quasi-channeling mode provide a large additional contribution near the atomic planes. Note that all the theory developed on the basis of the BBGKY chain is purely classical, while in the quantum theory the appearance of density peaks near the walls of the channel is attributed to the presence of particles on the above-barrier levels.

References

1. Bogolyubov N.N., Problems of dynamic theory in statistical physics, Moscow and Leningrad, Gostekhteoretizdat, 1949.
2. Kubo R., Statistical mechanical theory of irreversible processes. I. General theory and simple application to magnetic and conduction problems, *J. Phys. Soc. Jap.*, 1957, **12**, No. 6, 570–586.
3. Kubo R., Yokota M., Nakajima S., Statistical mechanical theory of irreversible processes. II. Response to thermal disturbance, Ibid, No. 11, 1203–1211.
4. Zubarev N.D., Statistical operator for non-stationary proceses, *Dokl. AN SSSR*, 1965, **164**, No. 3, 537–540.
5. Kalashnikov V.P., Nonequilibrium statistical operator in hot-electron theory, *Physica,* 1970, **48**, No. 1, 93–111.
6. Davis J.A., Friesen J., Mc.Intyre J.D., A radiochemical technique for studying range-energy relationships for heavy ions of keV energies in aluminium, *Canad. J. Chem.*, 1960, **38**, 1526–1532.
7. Robinson M.T., Oen O.S., Slowing down of energetic atoms in crystals, *Phys. Rev.*. 1963, **132**, 2385–2393.
8. Nelson R.S., Thompson M.W., The penetration of energetic ions through the open channels in a crystal lattice, *Philos. Mag.*, 1963, **8**, 1677–1689.
9. Erginsoy C., Wegner H.E., Gibson W.M., Anisotropic energy loss of light particles of MeV energies in thin silicon crystals, *Phys. Rev. Lett.*, 1964, **13**, 530–536.
10. Lindhard J., Influence of crystal lattice of motion of energetic charged particles, *Kgl. Dan. Vid. Sels. Mat.-Fys. Medd.*, 1965, **34**, No. 14, 3–63.
11. Kitagawa M., Ohtsuki Y.H., Modified dechanneling theory and diffusion coefficients, *Phys. Rev.*, 1973, **B8**, No. 7, 3117–3125.
12. Nitta H., Ohtsuki Y.H., Kubo R., Electronic diffusion coefficient for fast-ion dechanneling, *Phys. Rev.*, 1986, **34**, No. 11, 7549–7553.
13. Kagan Yu., Kononets Yu.V., Theory of the channeling effect, II.Effect of inelastic collisions, *Zh. Eksp. Teor. Fiz.*, 1973, **64**, No. 3, 1042–1064.
14. Kagan Yu., Kononets Yu.V., Theory of the channeling effect, III.Energy losses of fast particles, *Zh. Eksp. Teor. Fiz.*, 1974, **66**, No. 5, 1693–1711.
15. Zubarev, D.N., Kashlev Yu.A., Quantum theory of channelling: Generalised transfer equations, *Teor. Mat. Fiz.*, 1976, **29**, No. 3, 376–387.
16. Kashlev Yu.A., Quantum theory of channelling. Contribution of multiple scattering of channelled particles on the electrons to the dechannelling rate constant, *Teor. Mat. Fiz.*, 1979, **41**, No. 1, 89–101.
17. Kashlev Yu.A., Quantum theory of channelling, *Teor. Mat. Fiz.*, 1981, **47**, No. 1, 125–139.
18. Gombas P., Die Statistische Theorie des Atoms, Vienna, Springer, 1949.

19. Schiott H.E., Bonderup E., Andersen J.U., Esbensen H., Axial dechanneling I, A theoretical study, *Radiat. Effects,* 1972, **12**, 261–279.

20. Tikhonov V.I., Mironov, M.A., Markov processes, Moscow, Sov. radio. 1977.

21. Rozhkov V.V., Kinetic description of channelled particles, Vopr. At. Nauki Tekh., Fiz. Radiats. Povrezhd. Radiats. Materialoved., 1976, **1**, 5–7.

22. Calkin M.G., Nicholson P.J., Electrodynamics of a semiclassical free electron gas, *Rev. Mod. Phys.,* 1967, **39**, 361–372.

23. Kayser G., Statistical thermodynamics of nonequilibrium processes, Moscow, Mir, 1990.

24. Gardiner K.V., Stochastic or methods in natural sciences, Moscow, Mir, 1986.

25. Balescu R., Equilibrium and non-equilibrium statistical mechanics, volume 2, Moscow, Mir, 1978.

26. Kirkwood J.G., The statistical mechanical theory of transport processes. General theory, *J. Chem. Phys.,* 1946, **14**, No. 3, 180–201.

27. Suddaby A., Gray P., Relations between the friction constant and the force correlation integral in Browninan movement theory, *Proc. Phys. Soc.,* 1960, **75**, pt. 1, abstr. 481, 109–118.

28. Schwinger J., Brownian movement of quantum oscillator, *J. Math. Phys.,* 1961, **2**, No. 3, 407–422.

29. Kubo R., Problems of Statistical-mechanical theory of irreversible processes. Thermodynamics of irreversible processes, ed. D.N. Zubarev, Moscow, IL, 1962, 345–421.

30. Kashlev Yu.A., Local statistical channelling theory. Kinetic function in the conditions of interaction of fast particles with lattice atoms, *Teor. Matem. Fizika,* 2004, **140**, No. 1, 86–99.

31. Ellison J.A., Picraux S.T., Planar channeling spatial density under statistical equilibrium, *Phys. Rev.,* 1978, **B18**, No. 3, 1028–1038.

32. Ellison J.A., Continuum – model planar channeling and tangent squared potential, Ibid, No. 11, 5948–5962.

33. Tulinov A.F., Effect of the crystal lattice on some atomic and nuclear processes, *Usp. Fiz. Nauk,* 1965, **87**, No. 3, 585–598.

34. Chadderton L.I., Comments on the scattering of charged particles by single crystals, IV. Quasichanneling, flux peaking and atom location, *Rad. Effects,* 1975, **27**, 13–21.

35. Korta S., Hajduk R., Lekki J., et al., Experimental determination of the energy loss of protons channeled along the <111> axis through a silicon single crystal, *Phys. Status Solidi* (a), 1989, **113**, 295–306.

36. Oshiyama T., Dechanneling of fast charged particles from planar channel based on the stochastic theory, *J. Phys. Soc. Jap.,* 1980, **49**, No. 1, 290–298.

37. Appleton B.R., Erginsoy C., Gibson W.M., Channeling effects in the energy loss of 3–11 MeV protons in silicon and germanium single crystals, *Phys. Rev.,* 1967, **161**, No. 2, 330–349.

38. Nitta H., Semiclassical theory of dechanneling and diffusion coefficients, *Phys. Status Solidi (b),* 1985, **131**, 75–86.

39. Glauberman A.E., The kinetic theory of systems of interacting particles, *Zh. Eksp. Teor. Fiz.,* 1953, **25**, No. 3, 560–572.

40. Gurov K.P., Fundamentals of kinetic theory. Bogolyubov method, Nauka, Moscow, 1966.

41. Kashlev Yu.A., Sadykov, N.M., Statistical theory of channelling of fast particles, based on the local Boltzmann equation. Correlation matrix of

interactions and diffusion function of particles, *Teor. Mat. Fiz.*, 1997, **111**, No. 3, 423–496.

42. Ol'khovskii I.I., Sadykov N.M., Kinetic equation for the fast particles moving in a crystal under small angles in relation to crystallographic axes, *Izv. VUZ, Fizika*, 1977, **12**.

43. Green M.S., Markoff random processes and the statistical mechanics time-dependent phenomena, *J. Chem. Phys.*, 1952, **20**, No. 8, 1281–1295.

44. Zubarev D.N., Nonequilibrium statistical thermodynamics, Moscow, Nauka, 1971.

45. Gemmel D.S., Channeling and related effects in the motion of charged particles through crystals, *Rev. Mod. Phys.*, 1974, **46**, No. 1, 129–227.

46. Melvin J.D., Tombrello T.A., Energy loss of low energy protons channeling in silicon crystals, *Radiat. Effects*, 1975, **26**, 113–126.

47. Rozhkov V.V., Rutkevich P.B., The theory of thermal dechanneling from axial channels, *Phys. Status Solidi (b)*, 1981, **127**, 145–151.

48. Kashlev Yu.A., Shemaeva, I.N., in: Proc. 9th Conf. on Physics of interaction of charged particles with crystals, Moscow State University, Moscow, 1982, 203–207.

49. Hollinger H., Curtis C., Kinetic theory of dense gases, *J. Chem. Phys.*, 1960, **33**, No. 3, 1386–1402.

50. Green M.S., Markoff random processes and the statistical mechanics of time dependent phenomena, Ibid, 1952, **22**, No. 3, 381–413.

51. Abel F., Amsel G., Bruneaux M., et al, Backscattering study and theoretical investigation of planar-channeling processes, I. Experimental results, *Phys. Rev. B*, 1975, **12**, No. 11, 4617–4627.

52. Abel F., Amsel G., Bruneaux M., et al, Backscattering study and theoretical investigation of planar-channeling process, II. The unperturbed oscillator model, *Phys. Rev. B*, 1976, **13**, No. 3, 993–1005.

53. Rozhkov V.V., Theory of dechanneling, *Phys. Status Solidi (b)*, 1979, **96**, 463–468.

54. Ogletree T.W., Microsoft Windows XP, Dia Soft, 2002.

55. Davidson B.A., Feldman L.C., Bevk J., Mannaerts J.P., Statistical equilibrium in particle channeling, *Appl. Phys. Lett.*, 1987, **50**, No. 19, 135–137.

56. Kumakhov M.A., Spatial redistribution of the flux of charged particles in the crystal lattice, *Usp. Fiz.Nauk*, 1975, **115**, No.3, 427–464.

57. Kashlev Yu.A., Sadykov, N.M., The non-equilibrium statistical thermodynamics of channelled particles. Transition to the particles, moving at thermal velocities, *Teor. Mat. Fiz.*, 1998, **116**, No. 3, 146–160.

58. Zubarev D. N., Two-time Green functions in statistical physics, *Usp. Fiz. Nauk*, 1960, **71**, No. 1, 71–116.

59. Nagaoka Y., Theory of ultrasonic attenuation in metals, *Progr. Theor. Phys.*, 1961, **26**, No. 5, 590–610.

60. Kittel, Ch., Quantum theory of solids, Moscow, Nauka, 1967.

61. Kashlev Yu.A., Non-equilibrium statistical thermodynamics of dissipative processes in solids. Energy losses of channelled particles in the range of low velocities and adsorption of ultrasound by conduction electrons, *Teor. Mat. Fiz.*, 2004, **138**, No. 1, 144–156.

62. Zubarev D.N., Kalashnikov V.P., The derivation of time-irreversible generalized master equation, *Physica*, 1971, **56**, No. 3, 345–364.

63. Jaynes E.T., Informational theory and statistical mechanics, I, *Phys. Rev.*, 1957, **106**, No. 4, 620–630.

64. Jaynes E.T., Informational theory and statistical mechanics. II, Ibid, 1957, **108**, No. 2, 171–190.

65. Zubarev D.N., Time methods of statistical theory of non-equilibrium processes. Itogi Nauki Tekhniki. Sovrem. Probl. Matem., Moscow, VINITI, 1980, **15**, 131–226.

66. Bohr N., Passage of atomic particles through matter, Moscow, IL, 1950.

67. Kutcher G.J., Mittleman J.A., Atomic physics of channeled particles, *Phys. Rev.* A, 1975, **11**, No. 1, 125–134.

68. Lindhard J., Nielsen V., Scharff M., Thomsen P.V., Integral equations governing radiation effects, *Kdl. Dan. Vid. Sels. Mat.-Fys. Medd.*, 1963, **33**, No. 10, 1–42.

69. Lindhard J., Scharff M., Schiott H.E., Range concepts and heavy ion ranges, Ibid, 1963, **33**, No. 14, 1–42.

70. Feldman L.C., Appleton B.R., Multiple scattering and planar dechanneling in silicon and germanium, *Phys. Rev.* B, 1973, **8**, No. 3, 935–951.

71. Abel F., Amsel G., Bruneaux M., Cohen C., Oscillating trajectories in planar channeling studied by backscattering from iron crystals, *Phys. Letters, A*, 1972, **42**, No. 2, 165–166.

72. Ziegler J.F., Iafrate G.J., The stopping of energetic ions in solids, *Radiat. Effects*, 1980, **46**, 199–220.

73. Campisano S.U., Foti G., Grasso F.. et al, Depth dependence of angular dips: measurements and calculations for H^+ and D^+ in Si and Ge, Ibid, 1972, **13**, 23–31.

74. Matsunami N., Howe L.M., A diffusion calculation of axial dechanneling in Si and Ge, Ibid, 1980, **51**, 111–126.

75. Mori H.A., A quantum-statistical theory of transport processes. *J. Phys. Soc. Jap.*, 1956, **11**, 1029–1041.

76. Mori H., Correlation function method for transport phenomena, *Phys. Rev.*, 1959, **115**, No. 1, 298–312.

77. Mori H., Statistical-mechanical theory of transport in fluids, *Phys. Rev.*, 1958, **112**, No. 3, 1829–1842.

78. Mc Lennan J.A., The formal statistical theory of transport processes, *Adv. Chem. Phys.*, 1963, **5**, 261–371.

79. Rahman M., Quantum heat-bath theory of dechanneling, *Phys. Rev.*, B, 1995, **52**, No. 5, 3384–3396.

80. Mathar R.J., Posselt M., Electronic stopping of heavy ions in Koneko model, Ibid, 1995, **51**, No. 22, 15798–15807.

81. Cherchiniani K., Theory and applications of the Boltzmann equation, Moscow, Mir, 1978.

82. Prigogine I., Bak T.A., Diffusion and chemical reaction in one-dimensional condensed system, *J. Chem. Phys.*, 1959, **31**, No. 3, 1368–1370.

83. Katz M., Several probability problems of physics and mathematics, Moscow, Mir, 1965.

84. Pathak A.P., Interatomic potential and channeling, *Radiat. Effects*, 1976, **30**, 193–197.

85. Pathak A.P., Interatomic potential for atomic collisions in real solids, *Radiat. Effects Lett.*, 1979, **43**, 55–59.

86. Pathak A.P., Stopping power of solids in planar channeling, *Phys. Status Solidi, (b)*, 1978, **86**, 751–758.

87. Kashlev Yu.A., Quantum theory of channelling. the contribution of interaction of channelled particles with the lattice and kinetic coefficients, *Teor. Mat. Fiz.*, 1977, **30**, No. 1, 82–93.

88. Abrikosov A.A., Gor'kov, L.P., Dzyaloshinsky, I.E., The methods of the quantum field theory in statistical physics, Moscow, Fizmatgiz, 1962.

89. Jarvis O.N., Sherwood A.C., Whitehead C., Lucas M.W., Channeling of fast protons, deuterons and α particles, *Phys. Rev.*, B, 1979, **19**, No. 11, 5559–5577.

90. Van Hove L., Quantum-mechanical perturbations giving rise to statistical transport equation, *Physica*, 1955, **21**, 517–528.

91. Jackson D.P., Morgan D.V., Computer modelling of planar dechanneling, I: monoatomic lattices, *Radiat. Effects*, 1976, **28**, 5–13.

92. Ali S.P., Gallaher D.F., Electronic stopping power of channelled ions in model incorporating the Pauli priciple, *J. Phys.*, C, 1974, **7**, 2434–2446.

93. Vager Z., Gemmel D.S., Polarization induced in a solid by the passage of fast charged particles, *Phys. Rev. Lett.*, 1976, **17**, No. 20, 1352–1354.

94. Van Hove L., Quantum mechanical perturbation and kinetic equation., *Phys. Rev.*, 1957, **106**, No. 4, 874–889.

95. Bell R.P., Coefficient of permeability for parabolic potential, *Trans. Faraday Soc.*, 1959, **55**, No. 1, 1–4.

96. Barrett J., Potential and stopping-power information from planar channeling oscillations, *Phys. Rev.*, 1979, **20**, No. 9, 3535–3542.

97. Morita K., Dechanneling of heavy charged particles in diamond lattices, *Radiat. Effects*, 1971, **14**, 195–202.

98. Ohtsuki Y.H., Kitagawa M., Waho T., Omura T., Dechanneling theory with the Fokker–Plank equation and a modified diffusion coefficient, *Nucl. Instrum. and Methods*, 1976, **132**, 149–151.

99. Nitta H., Namiki S., Ohtsuki Y.H., Comparison between quantum and classical diffusion functions, *Phys. Lett.* A, 1988, **128**, No. 9, 501–502.

100. Dettmann K., Stopping power of fast channeled protons in Hartree–Fock approximation, *Ztschr. Phys.* A, 1975, **272**, 227–235.

101. Schtiller, W., Arrhenius equation and non-equilibrium kinetics, Moscow, Mir, 2000.

102. Datz S., Moak C.D., Noggle T.S.. et al, Potential-energy and differential-stopping-power functions from energy spectra of fast ions channeled in gold single crystals, *Phys. Rev.*, 1969, **179**, No. 2, 315–326.

103. Kubo R., Thermal ionization of captured electrons, Ibid, 1952, **86**, No. 6, 929–943.

104. Kubo R., Toyozawa Y., Application of the reproductive function method to irradiation and non-irradiation transitions of local electron in crystals, *Progr. Theor. Phys.*, 1955, **13**, No. 2, 160–182.

105. Solov'ev, G.S., Problems of the theory of the electronic structure and the energy of interaction of hydrogen in metals, Dissertation, Institute of Theoretical Physics, Kiev, 1976..

106. Bohm, D., Quantum theory, Moscow, Fizmatgiz, 1961.

107. Kashlev Yu.A., Non-equilibrium statistical thermodynamics of channelled particles. Transverse quasi-temperature, *Teor. Matem. Fizika*, 1995, **102**, No. 1, 106–118.

108. Kashlev Yu.A., Sadykov, N.M., Non-equilibrium statistical thermodynamics of channelled particles. Transition to particles moving at thermal velocities, *Teor. Matem. Fizika*, 1998, **116**, No. 3 442–455.

109. Brice D.K., Doyle B.L., Steady state hydrogen transport in solids exposed to fusion reactor plasmas, *J. Nucl. Materials,* 1984, **120**, 230–244.

110. Dumas H.S., Ellison J.A., Golse F., A mathematical theory of planar particle channeling in crystals, *Physica D*, 2000, **146**, 341–366.

111. Svare I., Hydrogen diffusion by tunnelling in palladium, *Physica B*, 1986, **141**, 271–276.

112. Esbensen H., et al, Random and channeled energy loss in thin germanium and silicon crystals for positive and negative 2–15 GeV/c pions, kaons and protons, *Phys. Rev.* B, 1978, **18**, No. 3, 1039–1053.

113. Derbenev Ya.S., Theory of electronic cooling, Dissertation, Novosibirsk, 1978.

114. Deans D., Vineyard D., Radiation effects in solids, Moscow, Mir, 1960.

115. Kelly R., Theory of thermal sputtering, *Radiat. Effects*, 1977, **32**, 91–100.

116. Kashlev Yu.A., Thermalization of displacement cascades in solids and the model of the thermal spike, *Teor. Matem. Fizika*, 2002, **130**, No. 1, 131–144.

117. Ikari H., Positronium in α-quartz. II. Possible effect of damping of positronium quasiparticle on the momentum distribution, *J. Phys. Soc. Jap.*, 1979, **46**, No. 1, 97–101.

118. Kohn W., Luttinger J., Quantum theory of electrical transfer phenomena, *Phys. Rev.*, 1957, **108**, No. 3, 590–617.

119. Zyryanov P.S., Klinger M.I., Quantum theory of electron transfer in crystalline semiconductors, Moscow, Nauka, 1976, 424–455.

120. Otsuki E.H., Interaction of charged particles with solids, Moscow, Mir, 1985, 144–169.

121. Mitchel K.H., Higher order transport equations for electrons and phonons derived from the Kubo formula, *Physica*, 1964, **30**, No. 12, 2194–2219.

122. Kinchin G.P., Piz R.S., Displacement of atoms in solids under the effect of radiation, *Usp. Fiz. Nauk*, 1956, **60**, No. 3, 590–615.

123. Kashlev Yu.A., Non-equilibrium statistical thermodynamics of cascade processes in solids. The quasi-temperature of cascade particles, *Teor. Mat. Fiz.*, 2001, **126**, No. 2, C. 311–324.

124. Morita K., Sizmann R., Dechanneling by interstitial atoms, *Radiat. Effects*, 1975, Vol. 24, 281–285.

125. Kitagawa M., Ohtsuki Y.H., Inelastic scattering of slow ions in channeling, *Phys. Rev.*, B, 1974, **9**, No. 11, 4719–4723.

126. Appleton B.R., Barrett J.H., Noggle T.S., Moak C.D., Orientation dependence of intensity and energy loss of hyperchanneled ions, *Radiat. Effects*, 1972, **13**, 171–181.

127. Robinson M.T., Deduction of interaction potentials from planar channeling experiments, *Phys. Rev.*, B, 1971, **4**, No. 5, 517–528.

128. Kashlev Yu.A. Non-equilibrium statistical theory of channeling. Energy–momentum balance equation and transverse quasi-temperature of a channeled particle subsystem, *Phys. Status Solidi (b)*, 1995, **190**, 379–390.

129. Barrett J.H., Appleton B.R., Noggle T.S., et al., Hyperchanneling, in: Proc. of Conf. Atomic collision in solids, N.Y., 1975, 645–668.

130. Goetzberger A., Fritzsche C.R., Diehi R., Superchanneling of accelerated

ions through crystals with large channels, *J. Appl. Phys.*, 1978, **49**, No. 9, 4956–4957.

131. Zimin N.I., Wedell R., Greschner B., Theory of planar channeling of relativistic positrons, *Phys. Status solidi (b)*, 1981, **105**, 257–263.

132. Kalashnikov, N.P., Remizovich, V.S., Ryazanov, M.I., Collisions of fast charged particles in solids, Moscow, Atomizdat, 1980.

133. Kelety T., Fundamentals of fermentation kinetics, Moscow, Mir, 1990.

134. Osipov A.I., Thermal dissociation of diatomic molecules at high temperatures, *Teor. Eksp. Khimiya*, 1966, **11**, No. 5, 649–657.

135. Campisano S.U., Foti G., Rimini E., et al, Temperature dependence of planar channeling in transmission experiments, *Nucl. Instr. and Methods*, 1976, **13**, 169–173.

136. Gartner K., Wesch W., Götz G., Temperature dependence of axial dechanneling by point defects, *Nucl. Instrum. and Methods Phys. Res.*, B, 1990, **48**, 192–196.

137. Goldman M., Spin temperature and nuclear magnetic resonance in solids, Moscow, Mir,1972.

138. Kittel Ch., Elementary statistical physics, Moscow, IL, 1960.

139. Gol'danskii V.I., *Physica*l chemistry of positron and positronium, Moscow, Nauka, 1968.

140. Ferrell R.A., Theory of positron annihilation in solids, *Rev. Mod. Phys.*, 1956, **28**, 308–337.

141. Akhiezer A.I., Berestetskiy V.B., Quantum electrodynamics, Moscow, Gostekhteoretizdat, 1953.

142. Hodges C.H., Scott M.J., Work functions for positrons in metals, *Phys. Rev. B*, 1973, **7**, No. 1, 73–79.

143. Brandt W., Paulin R., Positron diffusion in solids, Ibid, 1972, **5**, No. 7, 2430–2435.

144. Brandt W., Dupasquier A., Durr G., Positron annihilation in Ca-doped KCl temperature and magnetic field effects, Ibid, 1972, **6**, No. 8, 3156–3158.

145. Kashlev Yu.A., Distribution of flux density and the equation of energy balance of cascade particles in solids, *Teor. Mat. Fiz.*, 2000, **123**, No. 3, 485–499.

146. Brandt W., Positron dynamics in solids, *Appl. Phys.*, 1974, **5**, No. 1, 1–23.

147. Paynes D., Elementary excitations in solids, Moscow, Mir, 1965.

148. Carbotte J.P., Arora H.L., Thermalization time of positrons in metals, *Canad. J. Phys.*, 1967, **45**, 387–402.

149. Kadanoff L.P., Baym G., Quantum statistical mechanics, N.Y., Benjamin, 1962.

150. Seitz F., Koehler J.S. Displacement of atoms during irradiation, Solid state physics: Advances in research and applications, ed. F. Seitz, D. Turnbull, N.Y., Academic Press, 1956, **2**, 305–448.

151. Paulin R., Ripon R., Brandt W., Positron diffusion in metals, *Phys. Rev. Lett.*, 1973, **31**, No. 19, 1215–1218.

152. Yamada K., Sakurai A., Miyazima S., Hae Sun Hwang, Application of the orthogonality theorem to the motion of a charged particle in metals, II, *Prog. Theor. Phys.*, 1986, **75**, No. 5, 1030–1043.

153. Zeeger K., Physics of semiconductors, Moscow, Mir, 1977.

154. Sampaio A.J.C., Luzzi R., Relaxation kinetics in polar semiconductors, *J. Phys. Chem. Solids*, 1983, **44**, No. 6, 479–488.

155. Shank C.V., Auston D.H., Ippen E.P., Teschke O., Picosecond time resolved reflectivity of direct gap semiconductors, *Solid State Commun.*, 1978, **26**, 567–570.

156. Seymour R.J., Junnarkar M.R., Alfano R.R., Slowed picosecond kinetics of hot photogenerated carriers in GaAs, Ibid, 1982, **42**, No. 9, 657–660.

157. Suvorov A.L., Autoionic microscopy of radiation defects in metals, Moscow, Energoatomizdat, 1982.

158. Beavan L.A., Scanlan R.M., Seidman D.N., The defect structure of depleted zones in irradiated tungsten, *Acta Met.*, 1971, **19**, No. 12, 1339–1343.

159. Kirsanov V.V., Radiation physics of solids and reactor materials science, Moscow, Atomizdat, 1970.

160. Kirsanov V.V., Orlov A.N., Computer modelling of atomic configuration of metal defects, *Usp. Fiz. Nauk*, 1984, **142**, No. 2, 219– 264.

161. Zabolotnyi V.T., Starostin E.E., Fractal dimensionality of the longitudinal and transverse distribution of atomic displacements in cascades, *Fiz. Khim. Obrab. Mater.*, 2000, **3**, 5–8.

162. Coulter C., Parkin D.M., Damage energy functions in polyatomic materials, *J. Nucl. Mater.*, 1980, **88**, 249–260.

163. Semenov D.S., Analytical description of the characteristics of elastic collisions of atomic particles, *Poverkh. Fiz. Khim. Mekh.*, 1987 **6**, 41–44.

164. Thompson M., Defects and radiation damage in metals, Moscow, Mir, 1971.

165. Structure and radiation damage of constructional materials, Proceedings, edited by I.V. Gorynin, A.M. Parshin, Moscow, Metallurgiya, 1966.

166. Leiman K., Interaction of radiation with solids and formation of elementary defects, Moscow, Atomizdat, 1979.

167. Snyder W.S., Neufeld J., Vacancies and displacements in a solid resulting from heavy corpuscular radiation, *Phys. Rev.*, 1956, **103**, 862–864.

168. Wolfer W.G., Si-Ahmed A., On coefficient for bulk recombination, *J. Nucl. Mater.*, 1981, **99**, 117–123.

169. Kashlev Yu.A., The distribution of flux density and the energy balance equation of cascade particles in solids, *Teor. Mater. Fizika*, 2000, **123**, No. 3, 485–499.

170. Gann V.V., Marchenko I.V., Excitation of photons in the development of cascades of atom–atom collisions, Vopr. Atom. Nauki Tekh., Ser. Fiz. Radiats. Povrezhd. Radiats. Materialoved., 1987, **2**, No. 40, 61–62.

171. Lifshits I.M., Kaganov M.M., Tanatarov A.V., Theory of radiation changes in metals, *Atom. Energiya*, 1959, **6**, No. 4, 391–398.

172. Tetel'baum D.I., Semin Yu.A., Effect of the electronic subsystem and the work of elasticity forces on the parameters of thermal Spikes in ion bombardment of semiconductors, Proceedings of the 15th Conference on the Physics of interaction of charged particles with crystals, Moscow, Moscow State University, 1985, 101–102.

173. Kirkpatrick T.R., Cohen E.G.D., Dorfman J.R., Fluctuations in a nonequilibrium steady state: Basic equations, *Phys. Rev.*, A, 1982, **26**, 950–971.

174. Keizer J., Fluctuations, stability and generalized state functions at nonequilibrium steady states, *J. Chem. Phys.*, 1976, **65**, 4431–4444.

175. Bochkov G.N., Efremov G.F., Microscopic justification of non-linear fluctuation thermodynamics. Thermodynamics of irreversible processes, ed. A.I. Lopushanskaya, Moscow, Nauka, 1992, 21–30.

176. Shimomura Y., Fukushima H., Kami M., et al, Thermal stability of cascade

defects in fcc pure metals, *J. Nucl. Mater.*, 1986, **141–143**, 846–850.

177. Glansdorf P., Prigogine I., Thermodynamics theory of the structure, stability and fluctuations, Moscow, Mir, 1973.

178. Kramer J., Ross J., Stabilization of unstable states, relaxation and critical slowing down in a bistable system, *J. Chem. Phys.*, 1985, **83**, 6234–6241.

179. Sigmund P., Energy density and time constant of heavy ion-induced elastic collision spikes in solids, *Appl. Phys. Lett.*, 1974, **25**, No. 3, 169–171.

180. Protasov V.I., Chudinov V.G., Evolution of cascade region during the thermal spike, *Radiat. Effects*, 1982, **66**, 1–7.

181. Burenkov A.F., Komarov F.F., Kumakhov M.A., Temkin M.M., The spatial distribution of the energy generated in the cascade of atomic collisions in solids, Moscow, Energoatomizdat, Moscow,1985.

182. Thompson M.W., Nelson R.S., Evidence for heated spikes in bombarded gold from the energy spectrum of atoms ejected by 43 keV Ar⁺ and Xe⁺ ions, *Philos. Mag.*, 1962, **7**, No. 84, 2015–2026.

183. Zigmund P., Sputtering by ion bombardment: General theoretical considerations. Sputtering of solids by ion bombardment: Physical sputtering of single-element solids, ed. R. Berysh, Moscow, Mir, 1984, 23–98.

184. De la Rubia D., Irradiation-induced defect production in elemental metals and semiconductors: A review of recent molecular dynamics studies, *Ann. Rev. Mater. Sci.*, 1996, **26**, 613–649.

185. Bovbel' E.V., Val'dner V.O., Zabolotnyi V.T., Starostin E.E., The problem of the contribution of thermal spikes to the formation of the structure of the irradiated solid, *Fiz. Khim. Obrab. Mater.*, 2001, **1**, 5–7.

186. Gosar P., On the mobility of interstitials, *Nuovo Cim.*, 1964, **31**, No. 10, 781–797.

187. Kawasaki K., Time-dependent statistics of the Ising model: Coefficient of self-diffusion, *Phys. Rev.*, 1966, **150**, No. 1, 285–301.

188. Kashlev Yu.A., Three modes of diffusion migration of hydrogen atoms in metals, *Teor. Mat. Fiz.*, 2005, **145**, No. 2, 256–271.

189. Smirnov A.C., Theory of diffusion in interstitial alloys, Kiev, Naukova Dumka, 1982.

190. Kashlev Yu.A., Theory of incoherent hydrogen and deuterium diffusion in some b.c.c. nuclear materials, Diffusion processes in nuclear materials, ed. R.P. Agarwala, Amsterdam, Elsevier, 1992, 271–299.

191. Kashlev Yu.A., Solov'ev G.S., Statistical theory of diffusion, quantum states, configuration relief of the potential and to types of non-collinear end transition of the hydrogen atom in metals, *Teor. Matem. Fizika*, 1982, **50**, No. 1, 127–145.

192. Flynn G.P., Stoneham A.M., Quantum theory of diffusion with application to the light interstitials in metals, *Phys. Rev. B*, 1970, **1**, No. 10, 3966–3978.

193. Stoneham A.M., Theory of the diffusion of hydrogen in metals, *Ber. Bunsenges. Phys. Chem.*, 1972, **76**, No. 8, 816–822.

194. Stoneham A.M., Non-classical diffusion processes, *J. Nucl. Mater.*, 1978, **69**, 109–116.

195. Schober H.R., Stoneham A.M., Quantum diffusion of light interstitials, *Hyperfine Interaction*, 1986, **31**, 141–146.

196. Emin D., Baskes M.I., Wilson W.D., The diffusion of hydrogen and its isotopes in bcc metals, *Ibid*, 1979, **6**, 142–149.

197. Springer T., Investigation of the vibrational spectrum of hydrides of metals

by neutron spectroscopy, in: Hydrogen in metals, I. Basic properties, ed. G. Alefel'd, J. Völkl, Moscow, Mir, 1981, 94–125.

198.   Wipf H., Mageri A., Shapiro S.M., Satija S.K., Thomlinson W., Neutron spectroscopic evidence for tunneling states hydrogen in niobium, *Phys. Rev. Lett.*, 1981, **46**, No. 14, 947–955.

199.   Kashlev Yu.A., Theory of incoherent hydrogen diffusion in metals, *Physica A*, 1984, **129**, 184–200.

200.   Völkl J., Alefeld G., Diffusion of hydrogen in metals. Hydrogen in metals, I. Basic properties, ed. Völkl J., Alefeld G., Moscow, Mir, 1981, 379–408.

201.   Sussman J.A., A comprehensive quantum theory of diffusion, *Ann. Phys.*, 1971, **6**, No. 2, 135–156.

202.   Qi Zh., Volkl J., Wipf H., Hand D., Diffusion in Nb and Ta in presence of N interstitial impurites, *Scripta Met.*, 1982, **16**, No. 7, 859–864.

203.   Johnson H.H., Hydrogen in iron, *Met. Trans. B*, 1988, **19**, 691–707.

204.   Holstein T., Studies of polaron motion, I. The molecular crystal modes, *Ann. Phys.*, 1959, **8**, No. 2. 325–336.

205.   Stoneham A.M., Quantum diffusion in solids, *J. Chem. Soc. Faraday Trans.*, 1990, **86**, No. 8, 1215–1220.

206.   Klamt A., Teichler T., Quantum–mechanical calculation for hydrogen in niobium and tantalum, *Phys. Status Solidi (b)*, 1986, **134**, 533–543.

207.   Kehr K., Theory of hydrogen diffusion in metals. Hydrogen in metals, I. Basic properties, ed. I.M. Foelkel, G. Alefeld, Moscow, Mir, 1981, 238–273.

208.   Kashlev Yu.A., Mobility of particles in the one-dimensional lattice with several anharmonic effects taken into account, *Teor. Mat. Fiz.*, 1974, **19**, No. 3, 400–413.

Index

Milton Keynes UK
Ingram Content Group UK Ltd.
UKHW040446071024
449327UK00020B/1033